THE NOCTURNAL BRAIN

THE NOCTURNAL BRAIN

Nightmares, Neuroscience, and the Secret World of Sleep

GUY LESCHZINER

ST. MARTIN'S GRIFFIN
NEW YORK

Published in the United States by St. Martin's Griffin, an imprint of St. Martin's Publishing Group

 Printed in the United States of America. For information, address St. Martin's Publishing Group, 120 Broadway, New York, NY 10271.

www.stmartins.com

The Library of Congress has cataloged the hardcover edition as follows:

Names: Leschziner, Guy, author.
Title: The nocturnal brain : nightmares, neuroscience, and the secret world of sleep / Guy Leschziner.
Description: New York : St. Martin's Press, [2019] | Includes bibliographical references and index.
Identifiers: LCCN 2019005309 | ISBN 9781250202703 (hardcover) | ISBN 9781250202710 (ebook)
Subjects: LCSH: Sleep deprivation.
Classification: LCC RC548 .L47 2019 | DDC 616.8/498—dc23
LC record available at https://lccn.loc.gov/2019005309

ISBN 978-1-250-75697-8 (trade paperback)

Originally published in Great Britain by Simon & Schuster UK Ltd, a CBS company

First St. Martin's Griffin Edition: 2020

10 9 8 7 6

For Ava, Maya and Kavita. Also for Heinz, who always wanted to be a doctor – history intervened.

CONTENTS

	Introduction	1
1	Greenwich Mean Time	15
2	In the Still of the Night	36
3	Disney Was Right	58
4	Rumblings	76
5	The Sleep-Talking Bus Driver	94
6	Weak With Laughter	111
7	Buzzing Bees	142
8	Seized by the Throat	164
9	Floating Eyeballs	184
10	Jekyll and Hyde	198
11	The Waking Effects of Coffee	217
12	A Peculiar Fairy Tale	231
13	Inception	259
14	Losing Sleep	286

Epilogue: Some General Thoughts on Sleep 308

Appendix of Diagrams 316

Acknowledgements 323

Glossary 325

Further Reading 330

Index 343

THE NOCTURNAL BRAIN

INTRODUCTION

We think of sleep as a tranquil act, when our minds are stilled and our brains are quiet. The act of sleeping is a passive one, and is associated with a blissful unconsciousness and the delight of waking refreshed. The only awareness we might have of something happening in the night are the fragments of a dream. That is, at least, for most of us. But for many of the patients in my sleep clinic, their nights are anything but this. Rather, a night in the sleep laboratory, where I admit my patients to study their nocturnal behaviour, is punctuated by shouts, jerks, snores, twitches or even more dramatic goings-on, and the torture of poor or even no sleep at all.

The normal expectation of waking up feeling ready for the day ahead is rarely found among my patients, or indeed their partners. Their nights are tormented by a range of conditions, such as terrifying nocturnal hallucinations, sleep paralysis, acting out their dreams or debilitating insomnia. The array of activities in sleep reflects the spectrum of human behaviour in our waking lives. Sometimes these medical problems have a biological explanation, at other times a psychological one, and the focus of the clinical work that I and my colleagues do is to unravel the causes for their sleep disorders and attempt to find a treatment or cure.

For the past few years, I have seen hundreds of patients per year with sleep disorders, causing insomnia, profound excessive daytime sleepiness or bizarre and frightening experiences at night. My path to this work has been accidental. In keeping with most doctors of my generation, my exposure to the world of sleep during my medical degree was pretty much non-existent. I cannot recall a single moment of teaching that focused on sleep until well into my clinical training as a neurologist, almost a decade after I graduated. It was only by chance, when I opted for an intercalated degree in neuroscience as a nineteen-year-old, that I was asked to write an essay on the function of sleep. As a naive but intellectually curious teenager, I had assumed, like most people, that the function of sleep was to stop you feeling sleepy, and that assumption was born out of personal experience. I went to bed when sleepy, and when I woke up, that sleepiness had left me.

However, in preparation for that essay, I came across a paper co-authored by Francis Crick, one of the discoverers of the structure of DNA. Crick had in later life become increasingly fascinated by consciousness and neuroscience, in part driven by a sabbatical at the Salk Institute in San Diego, a world-leading centre in neuroscience research. In that paper, Crick and his colleague speculated as to the function of dreaming, which at that time was thought to exclusively happen during a stage of sleep known as rapid eye movement (REM) sleep. They argued that the function of dreaming, rather than representing a Freudian 'royal road to the unconscious', was a form of housekeeping for the brain. Dreaming, they postulated, acts to prune out connections between cells in the brain that have developed during the day, and constitutes a type of 'reverse learning' to get rid of useless information. The validity of this hypothesis remains controversial, but reading this paper was a light-bulb moment for an ignorant but interested medical student. The realisation that sleep not only had a purpose other than making you feel less sleepy, but was also a complex set of brain states, not simply a state of unconsciousness between going to bed and waking up, had a profound effect. It sparked my interest in sleep and its disorders, and has led me into this fascinating and often bizarre clinical realm of sleep medicine.

In this twilight world, glitches in the human brain result in striking and poorly understood conditions. All the more so as, in contrast to chest pain, headaches, skin rashes and more usual medical symptoms, these problems most often arise without any awareness, at a time when people's brains and minds are detached from their internal and external world.

In the pages that follow, I will introduce you to some of my patients who have been willing to share their stories. The tales of these individuals are dramatic, terrifying, illuminating, poignant and sometimes amusing. You will see how their disorders affect the lives of those around them, their relationships with partners and children, as well as their own.

So why is it that I'm writing about these patients? And, more importantly, why should you read about them? Many of the stories that follow are about patients with extreme sleep disorders, at the very limits of the spectrum of human experience, and it is by studying these extremes that we can learn about the less severe end of the spectrum; by understanding how these patients are affected by their sleep disorders, we come to know a little about how we ourselves are affected by our sleep. Many of these conditions are not rare, either: chronic insomnia affects one in ten adults; sleep apnoea about one in fifteen; and restless legs syndrome (RLS) about one in twenty. It is almost certain that anyone reading this book will either suffer from one or more of these disorders themselves or know someone close to them who does.

* * *

Doctors love stories; we love telling them and we love hearing them. We teach, learn and entertain each other with stories. In medical parlance, what a patient tells you in their own words is the history, the story, of their problem. As medical students and junior doctors, we acquire the skills to extract this history. Our medical journals and conferences are full of case histories, and it is precisely through the sharing of these stories that we circulate expertise and further our knowledge base.

I am a neurologist first and foremost, and the skills that I have learned through my neurological training are equally applicable in the practice of sleep medicine. As registrars (the equivalent of senior residents in the US) at the National Hospital for Neurology, Queen Square, in central London, we were exposed to a rite of passage: the venerated Gowers Round on a Thursday afternoon. Largely for teaching purposes, but also to provide some entertainment, this session takes place in a large lecture theatre with steep stalls. From the second row, where the neurology registrars sat, it felt a little like being in one of the amphitheatres of Rome – and we were about to be fed to the lions. The craftiest registrars among us would find a patient that urgently needed assessment on the wards so that they could creep into the rear of the auditorium late, along with the hordes of junior doctors, medical students and visiting neurologists from abroad. The most devious would arrange for a colleague to page them early in the proceedings so they could make a show of leaving to deal with 'an emergency' before sneaking in at the back of the lecture theatre later on.

The audience would await the sporting event with gleeful anticipation, while the registrars could only hope to survive the ordeal with a shred of dignity left intact. I have heard stories of colleagues vomiting in nervousness every Thursday lunchtime, others taking a beta-blocker pill to calm their anxiety prior to entering. For a painful ninety minutes, three cases would be presented. Usually the patients would be wheeled in at the front and the consultant chairing Gowers that day would grill the registrars on the cases, often exposing gaping holes in our knowledge under the glare of the 200 people sitting behind us.

After a particularly humiliating session, you would feel 400 eyes burning into the back of your head, as you wished that the earth would open up and swallow you whole. Some of my colleagues still talk about their most painful experiences twenty years later, such is the impact. (Even writing about it now I feel a slight flush, a delicate churning in the stomach . . .) As excruciating as these rounds were, they provided a fantastic opportunity to learn and to see conditions that you might never have heard of before, the knowledge perhaps reinforced by the sheer terror of the lesson. (I myself shall remember Triple-A syndrome and its association with neurological problems until my dying days, even though I have never heard it even mentioned since.)

While the fear of total humiliation in the Gowers Round sharpens the mind, it is hearing the complex stories of these patients that is its most valuable aspect. It is the patient history that physicians in general and neurologists in particular fixate upon, and the same can be said in sleep medicine. By far the most useful information when making or 'formulating' a diagnosis is the history, not the examination or the results of blood tests or scans. A man's recollection of some twitching in the left hand just before he fell and injured his head, suggesting a seizure arising in the right motor area of the brain, which leads to the diagnosis of a brain tumour; a young woman who reports visual loss spreading slowly over minutes across the visual field, confirming the visual aura of migraine – the spread of abnormal electrical activity associated with migraine headaches across the visual cortex – rather than an eye problem; the episode of dizziness several years previously that suggests that the woman sitting in front of you with a tingly hand may have multiple sclerosis rather than nerve impingement in the wrist; or the family history of imbalance, implying that the man with heavy alcohol use might have coordination difficulties due to a genetic disorder rather than as a result of his excessive drinking. The best neurologists I have worked with are the ones who have the patience and the ruthless determination to extract the full history, like a forensic FBI interrogator.

Such is the focus on the use of case histories for teaching that the case presentation is the standard way in which doctors are trained and their expertise is maintained. It enables us to 'experience' rare cases that we may at some point see in the future – hence the Gowers Round and its variations that exist in hospitals throughout the world.

During an admission into hospital, most patients are frustrated by the repetitive 'taking' of their history by medical students, tiers of junior doctors, generalists, specialty teams and consultants. The history is regurgitated and pored over again and again, and various aspects are repeatedly explored further. The impact of the condition on various aspects of one's life are surveyed but, as a rule, this facet of our patients' stories is something we do less well in the melee of a busy outpatient clinic, with the number of patients waiting outside our door ever increasing, under the irritated stares of people who have been waiting way beyond their appointment times. Our understanding of their relationship with their condition, how it affects their social or family life, and the minutiae of their complaints that are irrelevant to moving forward in the management of their illness are bystander casualties to efficiency. In reality, we simply try to extract all the necessary information to make a firm diagnosis and to formulate a treatment plan in the minimum possible time, so that we can move on to the next person.

As a schoolboy, I vividly remember picking up a copy of Oliver Sacks's book, *The Man Who Mistook His Wife for a Hat*. As I read these stories of a mariner unable to form new memories, of a man who could not recognise his own leg, of the woman who heard music as a result of epileptic seizures, I was gripped. But it was the context in which he put these symptoms, the impact on the lives of the human beings in front of him, that led to a deeper understanding of the nature of these conditions and how they affect us. And it was reading these stories that inspired my interest in neuroscience, and no doubt many of my colleagues too.

* * *

Neurologists are obsessed by 'lesions', the medical term for damage or injury. Whenever we assess a patient, we ask ourselves where the lesion is. We put together the symptoms and signs to 'localise the lesion', to identify its location in the nervous system. The damage may be due to a stroke, an injury, or a tumour. It may be visible to the naked eye or evident on a scan. It may be microscopic, only detectable after a biopsy or post-mortem. Or it might be transient, a 'lesion' that results from the temporary dysfunction of a small part of the nervous system due to an electrical aberration. But it is not just numbness of the arm or paralysis of the face that can be understood in terms of a lesion. Many of the sleep disorders that you will hear about in the following chapters are also a direct result of lesions.

Perhaps the most famous lesion of all in the world of neurology affected the brain of a man called Phineas Gage. Born in Grafton County, New Hampshire, Gage began working with explosives in his youth, perhaps on farms or nearby quarries. His introduction to blasting powder turned out to be very unfortunate for him, but very fortunate for modern neurology. At around 4.30 p.m. on 13 September 1848, near Cavendish, Vermont, while managing a work gang blasting rock to build a local railroad, the 25-year-old Gage was tamping explosive into a hole with a tamping iron, a long metal pole designed to pack explosive tightly. As he pushed it down, it must have sparked against the rock, igniting the explosive in the hole. The tamping iron flew out of the hole like a spear, whereupon it impaled Gage, entering the left side of his face, passing behind his left eye and smashing through the front of his brain and the top of his skull. The javelin-like rod landed some distance away, 'smeared in blood and brain'. Extraordinarily, after a brief convulsion, he sat up and was taken to the local doctor in an oxcart. According to this doctor's gruesome account,

> I first noticed the wound upon the head before I alighted from my carriage, the pulsations of the brain being very distinct. The top of the head appeared somewhat like an inverted funnel, as if some wedge-shaped body had passed from below upward. Mr Gage, during the time I was examining this wound, was relating the manner in which he was injured to the bystanders. I did not believe Mr Gage's statement at that time, but thought he was deceived. Mr Gage persisted in saying that the bar went through his head. Mr G. got up and vomited; the effort of

> vomiting pressed out about half a teacupful of the brain, which fell upon the floor.

Gage's survival, especially in the mid-nineteenth century, was truly remarkable. But even more so was the change that occurred in him after the accident. After a long convalescence, complicated by delirium, infection and coma, he finally made it to his parents' home some ten weeks later. It was not the same man who returned, however.

The details are scant, but prior to his accident he was described as hardworking, diligent and popular. His employers praised him as 'the most efficient and capable foreman in their employ'. After that dreadful accident, though, Harlow, one of his physicians, wrote,

> The equilibrium or balance, so to speak, between his intellectual faculties and animal propensities, seems to have been destroyed. He is fitful, irreverent, indulging at times in the grossest profanity (which was not previously his custom), manifesting but little deference for his fellows, impatient of restraint or advice when it conflicts with his desires, at times pertinaciously obstinate, yet capricious and vacillating, devising many plans of future operations, which are no sooner arranged than they are abandoned in turn for others appearing more feasible. A child in his intellectual capacity and manifestations, he has the animal passions of a strong man. Previous to his injury, although untrained in the schools, he possessed a well-balanced mind, and was looked upon by those who knew him as a shrewd, smart business man, very energetic and persistent in executing all his plans of operation. In this regard his mind was radically changed, so

> decidedly that his friends and acquaintances said he was 'no longer Gage'.

It seems that what had once been a pleasant, social man had been replaced with a belligerent, swearing and unpleasant character: 'He was gross, profane, coarse, and vulgar, to such a degree that his society was intolerable to decent people.' The story of Gage took on a life of its own, and was no doubt exaggerated and overblown with repetitive telling. In reality, it appears that in later life he was left less affected. But his was certainly one of the most famous historical cases of localisation, illustrating that different parts of the brain have different functions. Damage to the frontal lobes, whether through tumour, types of dementia or tamping iron, is known to cause personality change, and suggests that the frontal lobes have a fundamental role in our social behaviour and planning.

Correlating lesions to symptoms or signs, therefore, allows us to understand how our brains function, how they are organised and how our lives are determined by them. These lesions may be accidental or caused by disease. In animal experiments, they may be created by design. In clinical practice, we endeavour to characterise the location of the lesion in the nervous system. We attempt to form a unifying diagnosis, a single underlying cause to explain all the symptoms and examination findings.

In the world of sleep, however, this principle of Occam's razor – that the simplest explanation, a single diagnosis, should be sought to explain everything – does not always apply. Of course, in the neurology clinic, the explanation for a patient's migraine may be influenced by their stress levels or whether they have drunk alcohol, but for the most part this does not alter the diagnosis. In contrast, however, as anyone will testify, sleep is the absolute confluence of factors biological, social, environmental and psychological. Clearly anxiety may cause the tingling in your hands, and noise may worsen your migraine, but the link between your snoring, your work shift pattern, your noisy bedroom, your anxiety and your experience of sleep is so much more direct; these factors so much more fundamental to the difference between feeling rested and alert or exhausted beyond belief. Understanding all these aspects of your life is crucial to the evaluation of your sleep. But exploring all these facets can be a challenge in a thirty-minute consultation, especially when you are taking notes, struggling with the computer and dictating a letter at the same time.

Yet many of the sleep disorders that you will read about in the following chapters, like other neurological disorders, represent lesions of the nervous system – largely microscopic, transient or genetically determined, but lesions nonetheless. They are nature's experiments, giving us a window of opportunity to understand ourselves and help us identify how glitches in the brain's control of sleep can result in this huge array of phenomena. We will see how lesions of the brain result in uncontrollable sleep attacks, vivid dreams, hallucinations, sleep paralysis and collapses during the day. How abnormalities in the brainstem cause us to act out our dreams, and how genetic factors influence our ability to walk, eat, have sex or even ride a motorbike in our sleep. How chemical abnormalities in the nervous system can give rise to odd and distressing sensations at night. How our genes influence our body clock. And how seizures arising in our sleep can generate terrifying nocturnal experiences. Thus these phenomena can tell us about how our brains regulate our sleep, and how various aspects of our sleep are controlled.

Other patients in this book will illustrate how psychological or biological factors can influence sleep, causing debilitating insomnia, for instance, or sleep apnoea, where your breathing disrupts your sleep. One story in particular will demonstrate how a partner can have a huge impact on one's sleep. But even in these cases, when the cause is not related to damage in the nervous system, sleep itself is lesioned, disrupted or altered in some way. Through these case studies we also gain insight into the role of normal sleep in maintaining the brain – memory, mood, vigilance – through the impact of sleep deprivation or interference. These individuals provide windows into our understanding of the importance of sleep in the maintenance of physical, psychological and neurological health.

* * *

I am eager to introduce you to my patients and their stories, but before I do so, please forgive me a brief but important digression. To appreciate abnormal sleep, it is helpful to understand normal sleep. As we pass through life, our sleep changes, both in quantity and in quality. A newborn will sleep for two-thirds of the day, but by the time we are adults, we tend to sleep between roughly six and a half and eight and a half hours a night. Sleep is not a static state, however, and there are actually multiple stages involved.

As we first drift off, we enter into Stage 1 sleep, also known as drowsiness. The brain exhibits a quietening of normal waking electrical activity and the eyes slowly roll from side to side. As sleep progresses, we enter into Stage 2 sleep – light sleep – when the brain activity slows further. When we record the brainwaves during this stage, features called sleep spindles and K-complexes – transient alterations in the background brainwave rhythm not evident in wakefulness – become visible.

By the time we reach Stage 3 sleep – deep sleep – usually within about thirty minutes or so of drifting off, the brainwaves slow considerably but increase in size. This stage is therefore sometimes referred to as 'slow-wave' sleep. Stages 1 to 3 are considered non-rapid eye movement (non-REM) sleep, and it is only after sixty to seventy-five minutes or so that we enter into rapid eye movement (REM) sleep.

As we will see, in REM sleep, the eyes dart back and forth rapidly, the brainwaves look to be highly active – a little like being awake – and it is in this stage of sleep that we most obviously dream. As adults, over the course of the night we cycle through these various stages of sleep, usually four or five times, with the majority of deep, Stage 3 sleep in the first half of the night, and the majority of REM sleep in the second half.

As we age, the proportions of these various stages of sleep change. As newborns, we spend about half of our sleep in REM sleep, while in adults this ranges from 15–25 per cent, gradually falling as we approach old age. The proportion of Stage 3 sleep also changes, being roughly 15–25 per cent in adulthood, but dropping a little in the elderly, usually replaced by Stage 1 and 2 sleep. As we get older, the amount of wakefulness at night (very brief awakenings) increases too. As I will go on to show you, a complex system of brain nuclei, brain circuits and neurotransmitters regulate this biological process, controlling the initiation and termination of sleep, as well as the switch between non-REM and REM sleep.

There are two further processes that are important to grasp, since these mechanisms control the *drive* to sleep. The first is the homeostatic mechanism.

As anyone will know, the longer you have been awake, the stronger your drive to sleep. With prolonged wakefulness, levels of certain neurotransmitters that promote sleep build up, increasing sleepiness and thus promoting sleep onset. But the second potent force is that of the circadian clock, as we will go on to see.

Within us sits a timekeeper, an internal clock that co-ordinates our neurological and bodily functions with the external world. As we approach the dead of night, this clock exerts its strongest influence, compelling us to sleep, and in the daytime makes us feel more alert.

For the most part, these two mechanisms, the circadian and homeostatic, work in sync to ensure we sleep an appropriate amount at night and feel wide awake during the day. At least, they do when they are both working properly.

* * *

What I describe in the subsequent pages are patients I have seen over the years in the Sleep Disorders Centre, Guy's Hospital, and at London Bridge Hospital. I have been incredibly fortunate to know some of these people for many years and to have gained an insight into their conditions and their lives. For others, I have had opportunities to delve into their world more deeply, to meet them and their families in their homes, outside the constraints of the clinic, where time is less restricted and our discussions more leisurely. They have all consented and collaborated in the descriptions of their cases, ensuring accuracy and veracity. The only details changed are names where marked with an asterisk.

These patients illustrate the fundamental importance of sleep to our lives. And, as neurologist Oliver Sacks so aptly put it: 'In examining disease, we gain wisdom about anatomy and physiology and biology. In examining the person with disease, we gain wisdom about life.'

I

GREENWICH MEAN TIME

If you have ever been on a long-haul flight crossing time zones, the feeling of jet lag will be all too familiar. You know something is amiss: you feel sluggish and detached from your environment; the bright sunshine of your destination is discordant with your yearning to be tucked up in bed. There is the nausea of needing to stay awake when every fibre in your body craves sleep, or the incongruity of being wide awake at 2 a.m. while the world around you slumbers, and all you can think about is breakfast. Thankfully, your body soon adjusts, and within a few days you are back in tune with life around you. But imagine if that was how you felt all the time, that it was the reality of your daily life, and there was no hope of recovery.

I first meet Vincent, and his mother Dahlia, at Guy's Hospital. He is sixteen years old, and this particular clinic is specifically for teenagers transitioning from the sleep services in the children's hospital to the adult world. Typically, this clinic is full of children with narcolepsy or severe sleepwalking. But Vincent is not typical in this regard – or, indeed, any other. He is a shy and reserved teenager, not particularly tall but stocky and well-built. I learn that this is testament to his enthusiasm for boxing. Dahlia, in contrast, is bubbly and very talkative. Originally from South America, she speaks English fluently but with a strong accent and at a machine-gun pace. For the most part, Vincent sits there quietly as Dahlia tells me the story of the past few years, only interrupting when his frustration bubbles over. When he does talk, he is slow and hesitant; he occasionally finds it difficult to find his words.

Between them, they paint a picture of Vincent's life.

Vincent first became aware of some difficulties with sleep at around the age of nine or ten, but it was really only at the age of thirteen that his problems became much more evident. Dahlia thinks it started after Vincent had two operations on his hip, the second to remove metal plates inserted during the first procedure.

'Well, it was kind of gradual. At first I didn't really know what was happening,' Vincent tells me. He was initially finding it harder and harder to fall asleep, drifting off at three or four in the morning. 'The first time I properly realised it was a problem was when I was always trying to go to sleep, and then I started seeing the sun rise every time.'

It quickly got to the point where Vincent would be wanting to fall asleep at eleven in the morning and wake up at nine in the evening. Unsurprisingly, his schooling quickly began to suffer. 'I really missed a lot of school. At first I didn't want to tell anybody that I was having trouble sleeping, because they would just think that I'm lazy. So I just told them I was unwell a lot.'

For Dahlia, this time in their lives still stings. 'I started to notice when I was trying to wake him up to go to school that I could not wake him for love nor money. I would shake him, but just not be able to get him up. I was so confused because he had never been late for primary school. Never! I thought I was being judged as a mother. Possibly Vincent thought he was being judged as a student too. I got into so much trouble with his school. I was fined for Vincent's poor attendance!'

Vincent also recalls feeling judged: 'The school, my dad and friends found it hard to understand.' Some people, including his father, from whom Dahlia is separated, raised the likelihood that it was simply a case of a typical teenager oversleeping, or that it was psychosomatic. In fact, I think Vincent's father still considers this to be the case. On one occasion, I spoke to Dahlia on the telephone and I could hear him in the background, arguing with her that there was no medical issue.

Dahlia knew that there was more to it than teenage sleep patterns, however, and as Vincent's school attendance dropped further, she sought medical advice. Dahlia recalls taking Vincent to see their family doctor. 'We went maybe about seven or eight times, a few months apart, just to say Vincent has a problem with sleep. [We got] the usual recommendations – give him a hot milky drink before bedtime, no screens at night – all of that. Lavender oil . . .' she scoffs.

The problem nevertheless persisted, and eventually Vincent was referred to a paediatrician. It was at this point, some two years after he had realised he had a problem, that Vincent finally received a diagnosis: Vincent's internal body clock seemed to be set at the wrong time. Rather than being attuned to the world around him, he was told by doctors that his own body clock was running several hours later than everyone else's. He was diagnosed with delayed sleep phase syndrome.

* * *

We are all children of the sun. We are enthralled by it, and enslaved by it; we march to the beat of the sun's drum. Our sleep patterns are defined by the 24-hour rhythm of the rotation of the earth and our exposure to the sun's light. This makes total sense: to be awake and foraging for food when it is light and we can see prey and predators, and to sleep when it is dark and we are vulnerable to predation, seems crucial to our survival. It is not only our sleep that is defined by this rhythm, however.

Type 'circadian rhythm' – from the Latin for 'about a day' and the name for this 24-hour cycle – into PubMed, the most widely used search engine in the life sciences and medicine, and it will return over 70,000 hits – papers with titles ranging from 'Biological clocks and rhythms of anger and aggression' and 'Circadian regulation of kidney function' to 'Biological clocks: their relevance to immune-allergic disease'. Our 24-hour rhythm influences our brain, our gut, our kidneys, our liver and our hormones – every cell in our bodies. In fact, remove a cell, place it in a Petri dish, and it will demonstrate a 24-hour rhythm in some form or other. Indeed, 40 per cent of our genes that encode proteins are under the regulation of this circadian rhythm.

It is not simply a matter of exposure to light, though. The sun is not the metronome that keeps this rhythm going – at least not any more. Put humans in dim light, without any exposure to the rising and setting of the sun, and the rhythm will continue.

In the 1930s, Nathaniel Kleitman, one of the founding fathers of modern sleep science, experimented on himself and others in the depths of Mammoth Cave, Kentucky, the longest known cave system in the world. Deep underground, without light and without fluctuations in temperature and humidity, he tried to impose a 28-hour cycle, but found he could not. Even in the absence of the external cue of the sun's light, body temperature, sleep and other physiological parameters retain this 24-hour rhythm, implying that somewhere within us is a clock that keeps time.

It also seems that this clock is common to all life on this planet. Bacteria, single-cell organisms, plants, flies, fish and whales – they all have this endogenous clock. For some life-forms, the need for this clock is clear. But why should bacteria need to know what time it is, or indeed plants? Plants certainly need to know when the sun is shining, to know when to open their leaves and photosynthesise, but this does not need to be guided by an internal clock; simply detecting light would be enough. And why should fish living in cave systems, blind and not exposed to the light of the sun for thousands of generations, hold on to this clock? The fact that they do implies that this circadian rhythm is hardwired into the very essence of life, that since the existence of the last 'universal common ancestor', the very origin of all lifeforms on the planet, there has been an evolutionary pressure and natural selection acting to maintain this endogenous clock.

At the most simple end of life as we know it, bacteria and algae, it is difficult to know what this pressure might have been, however. It has been proposed that the origins may lie in a desire to avoid cell replication, which involves the copying of genes, during times of exposure to ultraviolet radiation, known to produce mutations. A more widely accepted hypothesis is that these rhythms evolved to control the production of genes that pre-empt and counteract daily fluctuations in oxygen levels and the damage that oxygen does. The circadian rhythm may in fact date back to the Great Oxygenation Event, approximately 2.45 billion years ago. This time period is defined by the evolution of bacteria called cyanobacteria, believed to have been the first microbes to achieve photosynthesis – the conversion of carbon dioxide to oxygen using energy from the sun's rays. At that time, atmospheric oxygen levels were low, and any free oxygen quickly became chemically bound to other substances. But the sudden rise in free atmospheric oxygen caused by cyanobacteria is thought to have provoked one of the largest mass extinctions in the history of the world, killing off most organisms for whom oxygen was highly toxic. Surviving organisms needed to develop mechanisms to protect themselves from the dangerous effects of free oxygen. It is thought that this need for protection resulted in the evolution of proteins called redox proteins, which mop up the toxic by-products of chemical reactions involving oxygen. The theory suggests that by predicting sunlight, and knowing when oxygen levels are going to rise, organisms can protect themselves from toxic damage, by generating these proteins at an appropriate time of day. But the truth is the origins of the circadian rhythm remain a mystery.

Any clock needs to be adjustable or reset, like a horologist tinkering with the pendulum of a grandfather clock to keep it running on time. The circadian rhythm, particularly for more complex organisms, needs to be tweaked according to the changing patterns of our seasons. Over the past few decades, our understanding of how this occurs has advanced. We are now aware of the influence of environmental cues or influences that gently nudge our circadian rhythms forward or back. These are termed *Zeitgebers* – 'givers of time' in German. Left to its own devices, the human circadian rhythm is set to 24.2 hours, and without *Zeitgebers* we would eventually find our internal clock drifting relative to the world around us. Our internal clock is sensitive to temperature, physical activity and eating, but by far the most potent *Zeitgeber* is light – particularly light at the blue end of the spectrum, like sunlight. While our circadian clock has proved itself independent of the sun, therefore, it is still its greatest influence.

The Royal Observatory, Greenwich, only a few minutes' train ride from the Sleep Disorders Centre, Guy's Hospital, sits atop a hill overlooking a large loop in the River Thames. From the thirtieth floor of the hospital, I can see the hill rising slowly towards south-east London, but cannot quite make out the building between the forest of ugly 1960s towers and new skyscrapers. On the roof of the observatory, a large metal mast with a weather vane on its tip juts into the typically grey London sky. On this mast, a large red ball, several feet in diameter, is impaled. Every day, at 12.55 p.m. Greenwich Mean Time in the winter, British Summer Time in the summer, the ball rises halfway up; then, at 12.58 p.m., ascends to the top. At 1 p.m. exactly, the time ball drops down the mast. In the present day, the area around the observatory is dominated by the skyscrapers of Canary Wharf, the main financial district of London, looming over the city from across the river. In the mid-nineteenth century, however, the Thames below would have been chock-full of sailing ships, ferrying the lifeblood of trade through the British Empire. Hundreds of telescopes would have been focused on the time ball of the observatory, waiting for the ball to drop. This would be the sailors' opportunity to reset the chronometer on board each ship to Greenwich Mean Time, crucial for the calculation of longitude on their journeys to the East Indies and beyond.

Like the chronometers on these ships, there are multiple clocks within the human body, but the seat of the master clock – the large red ball of the Royal Observatory – in humans, and indeed all vertebrates, is a tiny area of the brain called the suprachiasmatic nucleus. This tiny area, comprising a paltry few thousand neurones, sits in the hypothalamus, immediately above the optic chiasm, where the optic nerves carrying information from the eyes merge. This tiny nub of tissue is the control room for all circadian rhythms throughout the body, and destruction of the suprachiasmatic nucleus results in the loss of this rhythmicity.

Within the neurones of the suprachiasmatic nucleus, a complex dance occurs on a daily basis, with several genes with names like *CLOCK* and *Period* interacting with each other, feeding back to each other, conducting the ticking of our clock. But light, as a *Zeitgeber*, sways this dance, tweaking it forward or back. In the retina, at the back of the eye, in addition to the rod and cone cells responsible for converting light into vision, are cells known as retinal ganglion cells. A few of these cells have no contribution at all to vision. Their purpose is instead to conduct signals to the suprachiasmatic nucleus, through a direct projection called the retinohypothalamic tract. And it is through this pathway that light influences the rhythm in the suprachiasmatic nucleus, affecting the phase, the relationship of the 24-hour rhythm to the outside world, and the amplitude, the strength with which this rhythm runs. For people without any vision, the control of the circadian rhythm can be problematic, as we will see later.

* * *

The paediatrician's diagnosis in Vincent of delayed sleep phase syndrome is a common one. For those with this condition, their circadian rhythm runs behind that of the outside world. While most people want to go to sleep between 10 p.m. and midnight and wake between 6 and 8 a.m., people with delayed sleep phase syndrome may want to sleep at 3 a.m., sometimes as late as 7 a.m., and wake up seven or eight hours later. If they get this amount of sleep, then they feel fine. Unfortunately, life often gets in the way of sleep, and within the constraints of modern society, holding down a job or getting an education is difficult, if not impossible, on this sleep schedule.

To some extent, having a tendency to want to wake up early and go to bed early, or wake up late and go to bed late, is normal. There is a broad spectrum of chronotypes – a person's preference to go to sleep and wake up at a particular time. At the extremes of 'morningness' or 'eveningness' are those individuals known as 'morning larks' or 'evening owls'. People with delayed sleep phase syndrome can be considered extremes of the extreme, 'evening owls' whose circadian rhythm is so delayed that it has negative consequences on their life.

As with many features of our sleep, it appears that what chronotype we are is to some extent determined by our genes. Studies in twins or in families suggest that up to 50 per cent of our chronotype is under genetic control, and variants in the genes that regulate our circadian rhythm have been associated with both extreme 'eveningness' and extreme 'morningness'. In a familial form of what's known as 'advanced sleep phase syndrome', in which sufferers want to go to bed early in the evening and wake up extremely early in the morning, much rarer than delayed sleep phase syndrome, a mutation in one particular circadian gene, called '*PER*', has been identified. Furthermore, mutations in another one of these circadian genes, called '*DEC2*', seem to increase the amount of time we spend awake and reduce the amount of sleep required. For most people, however, it is not these few mutations that influence their wake/sleep pattern, but likely the cumulative effect of multiple milder variants in all of these genes.

Moreover, it appears that shifts in our chronotype also occur as the brain matures. Teenage circadian rhythms will typically shift later in the day, before then shifting back in adulthood. I can see this happening in my older daughter. Prising her out of bed in the morning is becoming increasingly difficult – as is getting her to go to sleep at a reasonable time at night. Undoubtedly this shift in the body clock seen in teenagers is compounded by the use of electronic gadgetry late in the evening. Being glued to your tablet, laptop or smartphone while in bed, as many teenagers are, provides a potent source of light to act as a *Zeitgeber* and makes this delay worse. This is a real problem. The consequence being that many teenagers, still needing to get up early to go to school, are sleep-deprived, and sleep deprivation is correlated with poorer performance at school as well as behavioural issues and anxiety. Individuals with delayed sleep phase syndrome, however, seem to be particularly sensitive to light exposure and its effects on the circadian rhythm. A burst of light in the evening seems to have a much greater delaying effect on the circadian clock in susceptible individuals than on average.

So, maybe the answer to Vincent's problem is as simple as cutting out the use of electronic devices at night. Or even wearing sunglasses in the evening to stop as much light as possible, especially blue light, from hitting his retinal ganglion cells. There is only one problem with this solution: Vincent does not actually *have* delayed sleep phase syndrome. What he has is much rarer.

If you listen carefully to his story, it is readily apparent, because Vincent does not want to go to bed at the same time every night (or day for that matter).

'Essentially my sleeping pattern shifts constantly, so my body wants to go to sleep an hour later every day,' Vincent says. 'So basically if I go to bed at 10 p.m. one day, I'll be naturally inclined to go to bed at 11 p.m. the next day, and so on.'

For Vincent, this constant shifting in his internal body clock means that bedtime, and by extension waking time, progresses by an hour a day. For a few days of every month, therefore, Vincent is synchronised with the world around him, but he soon shifts out of phase. 'For a week or so, I will be in social hours, but for the rest of the time, to different extents, I am out of sync.' At its worst, Vincent is essentially nocturnal and tells me that he can sometimes want to go to sleep at 11 a.m. and wake up at 9 or 10 p.m.

The impact of this shifting pattern is enormous. The result is that Vincent is often incredibly sleep-deprived. For most of the cycle that he shifts through, he finds it difficult to fall asleep at an appropriate time, but is forcing himself to get up in order to go to school. Some days, it is the equivalent of being rudely awakened at 2 or 3 a.m. and then being expected to pay attention in class at 4 or 5 a.m. Essentially, he is almost constantly jet-lagged.

Vincent says: 'When I'm in school, it can be very difficult to concentrate. One teacher noticed that my reading is particularly slow, and that it affects my processing skills. Sometimes it is almost impossible to stay awake and concentrate, so I could fall asleep during lessons.'

On one of the occasions we meet, it is about 5 p.m., but Vincent is in a phase when he wants to go to bed at 2 or 3 p.m. and wake up at midnight or 1 a.m. For Vincent, his brain is telling him he should be deeply asleep, and according to his body clock it is about 1 or 2 a.m. He struggles to string a sentence together, pausing constantly to find words, trying to get his thoughts in order. It reminds me of times as a junior doctor when I was on call for 24-hour shifts. I would be bleeped in the middle of the night and really have to pull myself together to give a sensible medical opinion. Vincent stumbles over his words: 'I just feel I'm behind the rest of the world right now. When I'm in sync with the world, I feel pretty good, because then I am able to be the best version of myself, the most articulate version of myself. Whereas right now, I'm not particularly.'

Dahlia's description of him when he is in sync and out of sync is striking:

> When Vincent is on a phase when he wants to sleep all day, when he is awake he is not himself. He looks tired,

> his responses are delayed, and he is mentally exhausted. Oh, when he is in sync with the world, when he normally wakes up at 6.30 or 7 in the morning, he's bright, he's like everyone else. He is passionate about his studies, he engages personally a lot more. He engages better with the world basically.

Unsurprisingly, Vincent's schooling has suffered terribly. 'It was getting very difficult to get in [to school] every day because I was constantly late, and teachers weren't being very understanding about the sleep disorder,' Vincent tells me. 'So after a while, I just dropped out because it was getting too difficult. It wasn't very sustainable.'

Dahlia is clearly bitter at Vincent's experience at school, and while she does not blame his teachers, she feels that there has been a lack of understanding and flexibility on their part regarding Vincent's medical issues.

It was not just Vincent's schooling that suffered; his social life was devastated too.

Dahlia says, 'I had to turn his friends away sometimes. When they came to visit him at, say, 7 p.m. to play PlayStation or something, Vincent would have been asleep since 5 p.m. So I had to say to his friends, "Oh, guess what, he's asleep!" But it is very strange for them, because a teenager never sleeps at 7 p.m.,' she laughs with a slightly bitter undertone.

* * *

Dahlia's determination to get to the bottom of Vincent's condition finally resulted in a referral to one of my sleep colleagues in the children's hospital. The history that Vincent and his mother give is absolutely typical of a condition called non-24-hour rhythm disorder, and this was confirmed using actigraphy – prolonged tracking of Vincent's sleep patterns using a wearable device, a medical version of some of the wrist activity trackers now widely available. Essentially, Vincent's circadian clock is running at twenty-five hours, rather than twenty-four. Somehow, Vincent's suprachiasmatic nucleus has become immune to or detached from the *Zeitgebers* – the external influences that normally nudge the clock to remain synchronised with the outside world.

In otherwise healthy individuals, non-24-hour rhythm disorder is really rather rare, but is much more common in people who are completely blind. It is easy to understand why. In the absence of any vision at all, that most important of influences on the circadian rhythm, light, is completely abolished as an input to the suprachiasmatic nucleus. The effects of other *Zeitgebers*, such as physical activity or eating, are magnified in its absence. The pathway from the retinal ganglion cells at the back of the eye via a dedicated bundle of fibres, the retinohypothalamic tract, is no longer intact. In fact, between half and two-thirds of patients who are unable to perceive any light have problems with sleep consistent with a circadian rhythm disorder. In one recent study, 40 per cent of totally blind individuals had a non-24-hour rhythm. In normally sighted individuals like Vincent, however, it is incredibly rare and poorly understood, though we do know that it typically starts in early teenagehood and is much more common in males.

We know that much of the influence of the intrinsic clock on the brain is mediated via a hormone called melatonin. This hormone is secreted by the pineal gland, a tiny pine-cone-shaped structure deep within the centre of the brain. René Descartes proposed this tiny area as the seat of the soul, though in reality its role is rather less glamorous, although still important. Under the influence of the suprachiasmatic nucleus, it churns out melatonin in a cyclical pattern.

For people with a normal sleep/wake cycle, melatonin levels rise in the early evening, stay elevated in the night and then drop back down a couple of hours before waking. The melatonin acts as a chemical signal to the rest of the brain that it is time to sleep, acting on melatonin receptors distributed very widely, not only in the brain, but also in multiple other tissues like the kidneys, gut, heart, lungs, skin and reproductive organs. So, by studying the rise and fall in melatonin levels in the blood, we can monitor someone's circadian rhythm, and the length of their cycle. But it is not quite as simple as that, because we know that a burst of bright light in the evening can suppress and delay this rise in melatonin before sleep. So, environmental factors can significantly alter the rise and fall of this hormone.

In order to understand someone's internal clock, they need to be kept in constant dim lighting conditions, bright enough to see but dark enough so as not to influence the pineal gland's secretion of melatonin. Looking at this pattern of melatonin in sighted individuals with non-24-hour rhythm disorder confirms that the sleep patterns seem to be internally driven, with an average cycle length of 25.2 hours – much longer than the 24.2 hours most people have. So maybe it is just having a cycle length that is so far off from the norm that is at least part of the problem. The effect of light and other *Zeitgebers* may simply not be strong enough to correct for such a big discrepancy.

Or it may be that there is just an insensitivity to the effects of light. Perhaps the suprachiasmatic nucleus is blind to the signals that the retinal ganglion cells send, like in patients who cannot see. In Vincent's case, his sleeping problems certainly get much worse in the winter months, and this might be directly associated with the lower intensity of light. A reduced effect of light on melatonin secretion has not been demonstrated in these patients, however, nor has a reduced sensitivity of the retinal ganglion cells ever been proven.

There seem to be some commonalities between people with non-24-hour rhythms and those with delayed sleep phase syndrome. In both, the natural rhythm is slightly longer than normal individuals, and analysis of the genes that define the circadian clock has shown variants in a gene called *PER3* to be associated with both patterns of sleep. So maybe it is the case that, if your rhythm is slightly longer than twenty-four hours, you have a tendency to run later, but eventually your rhythm is stabilised by the effects of *Zeitgebers*, causing delayed sleep phase syndrome. But if you run very long, and the drift is too great for the *Zeitgebers* to correct for, or the *Zeitgebers* simply don't work very well, then you end up free-running, like Vincent. This is still a hypothesis, and remains to be proven, but it is curious that there have been a few reports of non-24-hour rhythm disorder starting after the manipulation of sleep/wake patterns in patients with delayed sleep phase syndrome.

Chronotherapy involves delaying the bedtime by a certain number of hours every day in an effort to bring someone with delayed sleep phase syndrome back into sync. The rationale is that it is easier to stay awake for an additional few hours than to force your body to go to sleep earlier. By pushing your sleep pattern around the clock, you eventually get in line with everyone else. However, it seems that doing this can push the circadian clock to the limit, and in rare cases may result in the loss of control that one finds in non-24-hour rhythm disorder. In Vincent's case, perhaps it was the hip operations and the recovery period that were the initial disrupters of his sleep cycle.

* * *

Some of the effects of battling against your own body clock are readily apparent. The sleepiness or insomnia are obvious, as are the effects on cognitive ability, alertness and vigilance. The nurse at the workstation who briefly dozes off on their third night shift in a row is not a rare sight on the wards. It is not a reflection of laziness, but a direct function of their underlying circadian rhythm. The effects of the natural shift in the circadian rhythm in teenagers has even led some scientists and educationalists to propose that secondary school should start later in the day, to maximise the potential of pupils who are otherwise left sleep-deprived by waking earlier than their circadian rhythm dictates.

We are now beginning to understand that there are far-reaching and long-lasting implications of chronic disruption of the circadian clock, however. To comprehend the impact of this, studying the health of people who have been working shifts for long periods of time is a good place to start. For over twenty years now, we have been aware of some of the possible risks. A study in 1996 suggested higher levels of breast cancer in Norwegian radio and telegraph operators, and since then this finding has been reproduced several times. There is also evidence to point to an increased risk of colorectal and prostate cancers in shift-workers. The evidence is robust enough for the World Health Organization to add 'circadian disruption' to the list of probable carcinogens; and the Danish government to provide compensation to shift-workers with breast cancer. Moreover, it appears that shift work is also associated with gastrointestinal disorders, cardiovascular disease and diabetes.

So why should shift-workers have increased rates of certain cancers? One hypothesis is centred around the exposure to light at night. As we've discussed, light exposure at night suppresses the production of melatonin by the pineal gland, and it is argued that melatonin may have some anti-cancer activity above and beyond its role as a hormone – specifically, to absorb toxic by-products of oxygen metabolism that are thought to damage our DNA and predispose us to cancer. So, by regular exposure to light at night, perhaps we are lessening our resistance to cancer. This hypothesis is supported by the fact that people who are totally blind are less likely to develop breast cancer than normally sighted individuals, and in one experiment mutant mice predisposed to breast cancer were more likely to develop tumours when their circadian rhythms were disrupted. But there are lots of potential confounders.

We know that sleep deprivation in itself causes changes to appetite and promotes weight gain, a risk factor for breast cancer. And maybe shift work makes it more likely for you to take up an unhealthy lifestyle like smoking or exercising less. Also, very recently, a study has shown that even after three days of simulated shift work, the circadian clocks in the brain and in other organs become misaligned. The researchers found that markers of the brain's circadian clock in the suprachiasmatic nucleus remained relatively stable, but simultaneously there were dramatic changes in levels of the breakdown products of food. It could therefore be that this misalignment of brain rhythms and other 24-hour cycles in the body, usually tightly regulated, has fundamental consequences for how these products of food metabolism are processed, increasing our risk of diabetes, obesity and other medical issues. Furthermore, through this clash in the rhythms of various physiological processes in our bodies, our normal processes of cell replication and DNA repair may be impaired, giving rise to increased cancer risk. While the precise nature of the mechanisms underlying disrupted circadian rhythms and ill health remains poorly understood, certainly this association raises some very broad implications for us. Are we causing ourselves long-term harm through our exposure to light indoors and the use of electronic gadgets late into the night?

* * *

When my paediatric colleague told Vincent and his mother what the diagnosis was, there were mixed emotions. Vincent recalls feeling quite overwhelmed. 'It's a difficult concept that it's a chronic disorder, where no matter what happens you've got it for the rest of your life. That's quite a hard thing to get into your head.' But there was also a clear sense of relief. 'Before that [the diagnosis], I couldn't really be sure. Some people thought it might be psychosomatic.'

Dahlia tells me something very similar. She was expecting it. 'In my heart I knew. But when the diagnosis came, it was a relief too, because at least you know Vincent was not making it up. Vincent is not lazy. He was doing his very best. But on the other hand of course it's sad because it is something he has to cope with.'

The diagnosis has clearly had some benefits. Despite dropping out of school, Vincent has gone on to achieve excellent grades in his exams at secondary school. With a medical diagnosis in hand, he started at a school for children with special needs, and the flexibility has allowed him to achieve something close to his full potential. Vincent now attends a boxing academy, where he studies alongside training.

The diagnosis has also permitted treatment to commence.

For patients with delayed or advanced sleep phase syndrome, extreme 'evening owls' or 'morning larks' respectively, apart from trying to keep a strict sleeping regime, there are two major forms of treatment.

As well as being the chemical cue that the pineal gland

churns out to signal the chiming of the circadian clock, melatonin also directly influences that clock. Melatonin feeds back on the suprachiasmatic nucleus as well, and so is in itself a *Zeitgeber*. By giving people melatonin, the clock can also be shifted forward or back.

The other option is to manipulate light. Exposing people to very bright light, in the form of a lightbox, can also cause a shift. These lightboxes simulate natural sunlight, and are very rich in blue light, which seems to have the biggest effect on the retinal ganglion cells.

The timing of melatonin and light in relation to the underlying circadian clock is crucial, however. Depending on when in the circadian cycle melatonin or light is delivered, it can have very opposite effects. Exposing someone to bright light for an hour in the hour or two before natural bedtime can delay their bedtime by up to two hours; expose them to the same bright light in the morning after waking, and the bedtime will shift forward by about thirty minutes. Similarly, give someone melatonin in the early evening and they will go to sleep earlier; taken in the morning it will push bedtime back. In practice, we rarely give morning melatonin as it also potentially makes people drowsy, though there is some evidence that even small doses can cause a shift in the circadian clock without causing significant drowsiness.

Of course, for Vincent and others like him, there is no fixed rhythm that we can tailor melatonin and light timings to. But we can use these treatments to anchor his circadian rhythm. By giving his suprachiasmatic nucleus a regular evening dose of melatonin, and his retinal ganglion cells a dose of daytime bright light, his circadian rhythm is cajoled into staying a little more in sync with the outside world. And this regime, while not perfect, has made a significant improvement. Vincent still drifts a bit, particularly in the winter months, but treatment has made a big difference to his life.

'At the moment, I'm attending college and, so far, I am able to attend most of the time. And it's going okay. But I don't always feel 100 per cent.' Vincent tells me that he is currently managing to sleep at about 11 p.m. and get up at 6.30 a.m., and this cycle is fairly static. He has only missed a couple of days of school in the past few weeks. Every so often, however, despite our strategy to keep his rhythm regular, it still drifts. 'When my sleeping pattern goes off track and it's hard to bring it back, I'll stay awake for the next day and just not sleep at night. Then I'll fall asleep at a normal time again. That helps get me back into a pattern a bit quicker than waiting for several weeks. But it doesn't always work 100 per cent.'

I ask him about his boxing. 'I can be pretty inconsistent with my performance. So sometimes I can be much slower, or my reactions aren't quite so good [when I am out of sync]. I try to compensate by being faster and more powerful. But sometimes it's hard.'

We chat about what he thinks his future holds, what career he sees before him. 'I don't know. It's definitely going to be very difficult to fit in properly. There's not too much choice. Maybe self-employment or something like that. Something where I'm able to work on my own.'

* * *

For anyone who has done shift work, or is a regular traveller, Vincent's experiences will ring true. The disruption of your circadian cycle is disconcerting. I recall the drives into hospital at 3 a.m. on a Monday morning as a registrar, being called in to see someone with a stroke, feeling groggy and slightly nauseated, not thinking totally clearly. And even though I was passing through the streets of central London, one of the busiest cities on earth, I distinctly remember the feeling of being largely on my own – a strong sense of isolation; that I was not at one with the world. While the rest of the city was almost entirely tucked up in bed, here I was transgressing into a time of day that I had no business being awake in.

Ultimately, we are social beings. Although our circadian rhythms originated from our bacterial ancestors and have evolved to keep us awake in the sunlight and asleep in the dark, what I find remarkable is the importance they play in synchronising us as a social group, enabling people to live with similar rhythms: to eat at the same time; work at the same time; play at the same time; sleep at the same time.

This circadian clock knits our lives together as a species and as a society. And when one loses this clock, it sets us apart from the world around us; disconnects us from our family, friends and colleagues.

In Vincent's case, however, he cannot simply quit his job, or fly around the world a little less. For him, this is a constant and natural state of being. I am struck by the sense that it is this apartness without an end in sight, more than any other aspect of his condition, that is the most distressing. The loneliness of living one's life on a different rhythm to everyone else.

2

IN THE STILL OF THE NIGHT

Jackie is a grey-haired woman in her seventies, softly spoken, with a ready laugh. She speaks with what I take to be a gentle West Country burr, but which I later discover is a relic of her upbringing in Canada. She tells me of a time she went for a night-time ride on her motorbike.

She herself has no recollection of this moonlit ride, however, because she was asleep at the time. As well as her motorbike, she has also driven her car in her sleep. Without witnesses, she would be none the wiser to the fact that she is doing all this – managing to get dressed; ride or drive several miles; get undressed; and get back into bed – all without the slightest awareness that she has even left her bed. Disconcerting to the extreme, and almost implausible. Her doctor's response when she reported these events was to suggest admission to an in-patient psychiatric secure unit – not an appealing option for Jackie. And it is for that reason she finds herself sitting in my clinic.

Her letter of referral, from another sleep physician, was somewhat routine. 'Dear Guy,' it read. 'I would be grateful if you could see this lady, who was originally seen with a main complaint of sleepwalking, which she seems to be coping with quite reasonably, but it has been quite extreme.' There were some concerns about her breathing overnight, and some possible brainwave abnormalities on her sleep study, performed in her local hospital. But certainly nothing prepared me for the 'quite extreme' sleepwalking she went on to describe.

At first, I was incredulous. Many of the patients I have seen with this degree of 'sleepwalking' turned out to have a psychiatric or psychological root to their problems, like the woman I saw a few years ago who slit her throat and wrists with a kitchen knife, apparently in her sleep, or a young woman from Ireland who had been found eight miles away from home with her handbag and keys, having walked this distance without shoes. Jackie's rather matter-of-fact delivery of her tale, and initial apparent lack of concern over the degree of her night-time behaviour, did nothing to quell my cynicism. But, as I heard more about the background to her problems, I became more and more convinced about the nature of her sleep-motorbiking and sleep-driving.

Jackie's problems started decades before I met her. Born in the UK, she was brought up in Canada, and it was there that her sleepwalking first became apparent. 'I used to walk down the stairs to the lounge, open the door and stand in the doorway where my parents were,' she says, explaining her actions in the middle of the night. 'Well, it freaked Mother out, but Father just took my hand, took me back upstairs and put me to bed and that was it. But I've been doing it virtually since I could walk.'

It was when she joined the Brownies that her night-time antics became a problem. Needless to say, Jackie was definitely not the most popular girl to share a tent with. In the middle of the Canadian wilderness, her sleep activities were particularly unfortunate. 'I used to make this sort of growling,' she says. 'But it wasn't a quiet sort of growl. I think they thought a bear was coming after them. I growled so much and so loud that it frightened them, so they wouldn't have me around.' She was also a handful for the adults supervising the trips. 'I'd get up in the middle of the night and I'd walk down to the river. I'd walk into the woods. They couldn't cope with me, so I had to be picked up and taken back home again.' She laughs as she tells me these stories, but I imagine that the impact of this on her as a child was terrible, and perhaps caused her to be a little socially isolated.

For anyone who is a parent, some of Jackie's nocturnal behaviours will sound very familiar. Sleepwalking and related issues are incredibly common in childhood. Most traumatising, for parents rather than the children, are sleep terrors, when children will scream and cry inconsolably in the middle of the night and then go back to sleep, subsequently waking without any recollection. These conditions are termed non-REM parasomnias, as they arise from non-dreaming, very deep sleep. Trying to wake a child from deep sleep is highly likely to generate some sleep-talking or even sleepwalking.

What is less common is the persistence of sleepwalking into adulthood, which occurs in about 1–2 per cent of people. Jackie is one of these people. Her episodes persisted through to early adulthood, after her move back to the UK. It was shortly after her return that her sleepwalking took a twist. She was lodging with an elderly woman when one morning she came down for breakfast. Her landlady greeted her with a confusing question. 'She said, "Where did you go last night?"' Jackie tells me. She denied having gone anywhere. 'Well, you went out on your motorbike,' her landlady replied. Jackie recalls being completely shocked, and initially perplexed. It is easy to imagine her incomprehension at what she had just been told. From her perspective, she had simply gone to bed that night as usual, and had woken up as usual. She immediately asked if she had been wearing her helmet. 'Oh, yes, you clomped down the stairs and you had your helmet and you went out,' her landlady said, adding that she had been gone for about twenty minutes. There were no other clues, as she had returned the motorbike to exactly the same place she had left it.

After a few more night rides, Jackie gave her motorbike keys to her landlady for safekeeping, and later sold it. She still misses her BSA 250. 'Brilliant bike! You can hear that coming for miles.' I tell her it is surprising that it did not wake her up. 'It is, isn't it?' she says.

* * *

So how does medical science explain what is happening in Jackie's case, these complex behaviours such as walking, growling or even motorbiking in deep sleep? We have known for years that certain animals like dolphins, seals and birds can sleep with one half of their brain at a time, allowing them to swim or fly while sleeping, what is termed 'uni-hemispheric sleep'. Aquatic mammals obviously need to be able to swim and surface to breathe, but like us they must also sleep, so this neat evolutionary trick prevents them from drowning while carrying out these necessary functions. It also stresses the importance of deep sleep from an evolutionary perspective: if deep sleep served little useful purpose, why would this uni-hemispheric sleep be necessary?

In humans, however, uni-hemispheric sleep does not exist. We used to think of sleep being an 'on' or 'off' brain state: either you are awake or you are asleep; there is nothing in between. But, in recent years, we have learned that this is not the case. Deep sleep and full wakefulness lie at the extremes of a spectrum, and, implausible as it may sound, it is possible for us to be in both states at the same time.

When monitoring brainwaves using electrodes attached to the scalp, as we do when we undertake sleep studies on patients, the hallmark of deep, non-REM sleep is of synchronous electrical activity throughout the brain – slow waves of high amplitude termed 'delta' waves. But in these sleepwalking events, the picture can be very different. Mixed in with these slow waves, sometimes we see brainwave activity that looks very much like the activity of a brain that is wide awake, suggesting that wake and sleep are occurring at the same time. Using scalp electrodes really only gives us a limited glimpse of what is happening within the brain, though, like looking into a room through a keyhole. It is impossible to get the whole picture. This technique only provides information about what is happening close to the surface of the brain; nothing about what is happening in the brain's core.

But there are other ways of looking at the brain. In 2000, Swiss researchers managed to capture brain activity during sleepwalking using a technique called SPECT. This involves the injection of a radionuclide, a radioactively labelled chemical, and rather than telling us about the structure of an organ, it tells us about its activity. This 'dye', the injected radionuclide, concentrates in areas where the blood flow is greatest, which equates to areas of greatest metabolic activity, i.e. those tissues with the highest demand for oxygen. In a feat of timing, these researchers managed to inject the radionuclide within twenty-four seconds of a sleepwalking event starting, in a sixteen-year-old boy who was known to sleepwalk several times a week. Even more impressively, they managed to do all this with him lying in a SPECT scanner, to detect where in the brain the radioactive substance was most concentrated. By comparing this scan to a scan performed in deep sleep, what they found was remarkable. During the sleepwalking episode, a deep area of the brain called the posterior cingulate cortex was found to be very active, while another area, the frontoparietal cortex, showed significantly reduced activity compared to wakefulness. Essentially, what they found was that small areas of the brain were awake, while other areas of the brain remained asleep. Those areas with increased activity, the cingulate cortex in particular, are involved in the control of behaviour associated with strong emotions. In contrast, the frontoparietal cortices, especially an area called the prefrontal cortex, where there was reduced activity, are involved in planning, rational thinking and personality. This pattern in sleepwalking, particularly in the sixteen-year-old, in whom sleepwalking often had elements of fear associated with it, makes perfect sense. During these events, the part of his brain involved in strong emotion was in overdrive, almost awake, while the part of his brain involved in logic, in personality, in planning his actions, remained deep asleep. It seems that this dual brain state of simultaneous wake and sleep explains these phenomena of complex activities: the ability to interact with the world around you, without the ability to think rationally, as you would when you are awake.

Serendipitously, an Italian group also captured sleepwalking in a twenty-year-old man, this time with electrodes implanted into the brain during monitoring for possible surgery for epilepsy. For patients suffering from epilepsy that does not respond to medication, surgical removal of the area of the brain triggering seizures can sometimes be an option, but relies upon the accurate identification of precisely where the seizures are coming from. Because of the limitations of recording electrical activity from the scalp, small wires are inserted into the skull, usually over the surface of the brain but sometimes with electrodes inserted deep into the brain substance. In this case, the poor young man had suffered from epileptic seizures since the age of seven, having had meningitis as an infant. Unfortunately for him, although fortunately from our perspective, he had also experienced sleepwalking since even before his meningitis diagnosis, and during the course of his study, the patient experienced both seizures and sleepwalking. With electrodes recording directly from within the brain itself, on one occasion during deep sleep he turned in his bed, extended his arms as if to embrace someone, gave a kiss, and muttered a few words before going back to sleep. During this sleep-kissing episode, electrical activity in motor and cingulate areas once again showed waking activity, while other areas remained in very deep sleep. This study would appear to confirm the conclusions of the SPECT study performed a few years earlier, proving that those earlier imaging findings were truly a reflection of sleep state rather than simply a result of blood-flow changes in the brain.

So it seems that sleep, rather than being a global phenomenon affecting the whole brain, is a local occurrence. The brain does not act as a whole, and in these unusual cases, different parts of the brain can simultaneously exist in different states of wakefulness or sleep. In a similar way to a dolphin sleeping with only one hemisphere, it appears that this can happen in the human brain, albeit on a more local level.

* * *

Alex is another of my patients, now in his twenties, but a regular sleepwalker since childhood. He is tall, well-spoken, with long hair and an earring. He lives in a shared house in a rapidly gentrifying area of south London. He is working for a charity at present, but intends to travel the world. When I first meet him, he attends clinic with his mother. The referral letter states: 'He is already taking common-sense steps to do with his parasomnia, but if anything things are becoming more frequent, and the risk of him getting into trouble is increasing. I would therefore be grateful if you could arrange to see him before he succumbs to some unfortunate mishap.'

His mother is understandably concerned, but Alex, like Jackie, is pretty relaxed about it all. It has been part of his life for as long as he can remember. At boarding school, his sleep-walking was the subject of initial alarm. 'When I first started at school, we were in a dorm of sixteen,' says Alex, 'and when everyone first knew about it, I fell asleep with my earphones round my neck and I woke and I thought there was a rat sitting on my chest, so I screamed "Rat!", and that woke everyone up in the dorm room. And obviously there wasn't a rat, but people were terrified.'

Terror was soon replaced by amusement, though. Speaking of his roommate, Alex explains: 'One time he woke up and apparently I was on my desk shouting that there was a tiny gazelle running round trying to eat my feet. At first he was really terrified, and then as it happened more and more he found the funny side of it.'

Over the years, the spectrum of Alex's non-REM parasomnias has broadened. Some of his events are simply amusing. On one occasion, having been on a night out, he got to his friend's house at about 2 a.m. before going to sleep. The next thing he knew, he was waking up in the neighbour's front garden at 5 a.m., in just his underwear. He was being shouted at by a man through an open window; apparently he had tried to get into the neighbour's house, waking up his wife and baby in the process. On another occasion, a different friend was woken by Alex at 3 a.m., crouched at the end of the bed, ordering a pizza. He was using his friend's shoe as a telephone. His current flat-mates also tell countless anecdotes about Alex's exploits, which he listens to with a resigned grin. I suspect over the years he has got used to this. 'There was one time we'd been at the pub one night, a group of us,' says Gareth, a housemate of Alex. They had all retired to their respective bedrooms.

> Then, an hour later, he just comes into my room, nonchalant, [and] gets into bed with me. And I was awake at the time, but was a bit tired and I was nudging him saying: 'What's going on?', and he was having absolutely none of it, and then in the morning he wakes up and says: 'Why am I in your bed?', as if I'm supposed to have the answers. And the same thing happened another time. I woke up and there he was again. He was pretty sheepish and I said: 'You're naked under there, aren't you?', and he was like, 'Yeah, yes I am.' The next morning someone said: 'You know, I actually went into your room for a chat and I saw someone else in there. So who's the lucky girl?', and I had to admit, 'Well, actually, that was Alex.'

Another housemate reports further events.

> I used to live in the room next to his, and [one night] I heard a great crashing and quite a lot of swear words. I ran in to make sure that everything was okay and he was lying on the floor – I don't know why he still continues to sleep fully nude considering all these things that he does. He said that he thought that a train was coming at him, so he leapt up and bounced off his wall. He swore quite a bit because he thought the train was still coming at him. So he then leapt onto his desk which then crashed on top of him. So I ran into the room to make sure he was okay just to find him very, very confused. I now live in the room underneath him and I was woken up three times in the

night because he believed that the circus was in town, but the manager of the circus needed to use our bathroom, so he kept going and opening the door and muttering nonsense to someone who wasn't there and so I had to walk him back up to bed.

In contrast, some of Alex's parasomnias have a darker side. The more comical episodes are often not remembered by Alex, and he is frequently only aware of them when his friends regale him with tales of his night-time antics in the pub. But it is those events associated with strong emotion, typically fear or anger, that Alex is more conscious of. 'I remember much more the emotionally draining ones,' he explains. 'When I sleepwalk, I have no memory of those whatsoever, so that's not when I feel panicked.' The more terrifying events that disturb Alex's sleep are often scenarios charged with the need to fight or run away – the so-called 'fright-fight-flight' response, mediated by adrenaline and the autonomic nervous system, linked to physical functions related to these responses to danger. Alex recalls that many episodes of his non-REM parasomnias are related to a snake in the bed, a nuclear bomb about to go off next door, or impending peril. 'I fractured my finger once trying to save a girl from drowning. I leapt and tried to, like, catch her, but it was just the cupboard I lunged at and so I caught my finger quite badly. And I've tried to climb out of a window a few times, but then at the last second I've sort of stopped myself from doing it.'

His ex-girlfriend Katie confirms this. She says that many of his events relate to natural disasters, and she has frequently been woken by him trying to drag her out of bed. Katie says:

I wake up and Alex is panicked and he has gotten up . . .

> He is running round the room trying to get out. And then obviously he's aware that I'm there and he tries to pick me up and sort of drag me out of bed as well. And at that point I'm already sort of awake, calling: "Alex, Alex!" It takes me quite a long time to get through to him and then eventually he wakes up and that's it.

His housemate Gareth reports another occasion when Alex injured himself. 'Two of his friends were sleeping on mattresses on the floor and he believed that the rotary blades of a helicopter were breaking off while they were spinning and flying at him, so he was dodging and diving to move from them,' Gareth tells me. 'And as the final one came at him, he completely swan-dived onto what he thought was grass but was actually our two friends on their mattress. They were woken up at 4 a.m. by a quite large Alex landing on them saying, "Oh, God, did you see that helicopter?"'

* * *

Many of Alex's events are similar to Jackie's. His sleepwalking, like Jackie's sleep-motorbiking, involves complex activities and interaction with his environment – picking up a shoe to call for a takeaway pizza, getting into bed with his flatmates, and on one occasion finding tumblers, filling them with water and placing them next to the bed of each of his sleeping housemates in the middle of the night – an act of unconscious kindness. These events, as with Jackie, happen without any awareness, with no memory, and perhaps importantly are fairly neutral in terms of emotional content. These types of non-REM parasomnias represent a classical view of this condition, that people with this type of problem generally have no recollection at all of these events. But it is clear that Alex is certainly aware of some of his events. It is also clear that Alex is having dreams of a sort, despite me telling you that these phenomena emanate from non-REM sleep as opposed to REM or dreaming sleep. In fact, in recent years, we have begun to understand that previous notions of REM sleep being the stage of sleep exclusively in which we dream is not correct. It seems that dreaming in non-REM sleep is not unusual at all, although the content of the dreams is different.

Individuals woken from REM sleep often describe dreams with a narrative structure, a story evolving like the plot of a book or a film. In contrast, the dreaming of non-REM sleep often contains only simple visual imagery, like animals, people or inanimate objects. When mixed in with strong emotional stimulation, these images are often scary or even terrifying – spikes crashing down from the ceiling, insects on the walls, or snakes in the bed. And it appears that this surge of adrenaline mediated by the autonomic nervous system means it is more likely for people in this sort of episode to wake up, sometimes partially but sometimes fully, and to remember their experiences, like Alex's more fear-inducing events.

And the origin of this 'fright-fight-flight' response? It seems that one of the areas identified in neuroscientific studies of sleepwalkers as being particularly active, the cingulate cortex, part of the limbic system, is the neurological seat of this experience. Triggering this area, either directly using electrical stimulation or when epileptic seizures arise in this region spontaneously, results in the adrenaline surge that apparently feeds into Alex's more distressing sleep terrors.

* * *

So it seems that sleepwalking, and conditions within the spectrum of non-REM parasomnias, represent a battle between wake and sleep in various parts of the brain. At one end of the spectrum, there is simple sleep-talking, sleepwalking or sleep-sex (see Chapter 10), when there is no awareness, no emotion, but the ability to speak in fully formed sentences or to move. Very few parts of the brain are awake, perhaps only those that allow people to see, to move or to talk. The areas of the brain responsible for rational thought and memory certainly seem to be in deep slumber. At the other end of the spectrum are sleep terrors such as Alex's, where the emotional stimulus is so intense that it drives the person to be almost fully awake, whereby the majority of the brain is fully functioning, with the exception of a bit of rational thinking perhaps. In children, these sleep terrors are generally not remembered, which may well simply be a function of how deeply young children sleep and the level of stimulation required to fully wake a child in deep, non-REM sleep.

And what about Jackie's sleep-motorbiking? Where does that fit in? Clearly, during these activities, she must have the ability to see, and probably to hear. She can get dressed, pop her helmet on, grab the keys, change gears, avoid a crash, and navigate home, then get undressed and back into bed. In fact, the only areas of her brain clearly not functioning properly are those involved in memory, hence her lack of awareness, and in rational thinking, since who would rationally get up in the middle of the night, go for an aimless motorbike ride and then go back to sleep? So is this kind of behaviour really sleep, with a bit of wake, or is Jackie essentially awake, with a small area of her brain asleep?

* * *

Having handed her motorbike keys to her landlady, and eventually selling her motorbike, Jackie thought she had dealt with her sleep issues quite successfully. She had subsequently always lived on her own, and until recently has always been single, and had not had any further problems with her sleepwalking.

For the past few years, she has been living in an apartment in a block directly on the seafront in Seaford, a small peaceful town on the coast of Sussex. The pace of life suits her, and she enjoys volunteering, keeping the coastal paths clear of shrubs and in good condition. 'Seaford is a wonderful place to live,' she tells me. 'The sea cliffs are right up there, and over in that direction are the Downs,' she points out to me as we stand on the pebble beach in front of her apartment. With the waves gently lapping the shingle and the seagulls crying overhead, this quiet little town strikes me as an unlikely location for such bizarre happenings at night. She has good relations with her neighbours, and her block of flats is a friendly place. Few of her neighbours know of her sleepwalking past, except for one friend, with whom she went on a cruise. Jackie would wander the ship at night, and eventually took to asking the cabin crew to take her key card away at night and slip it back under the door at 6 a.m.

* * *

One morning, a few years ago, she came out of her apartment and bumped into a couple living in the same block. To her surprise, they asked her what she had been doing at 1.30 or 2 a.m. 'I said, "I was asleep." They said, "No, no, you were driving out [of the car park] as we were coming home. We were coming back late from a show quite a way away and you were driving out."' Initially, a puzzled Jackie denied going anywhere, but soon it dawned on her what might have happened.

'Okay, where did I go?' she asked.

'Oh, well, you turned left out of the road and you went down towards the promenade.'

'Well, okay. Did you see me return at all?' she pressed.

'No, we went straight in and went to bed. We just wondered where you were going.'

'I have no idea,' Jackie replied.

To her alarm, Jackie realised that as she had previously been sleep-motorbiking, she was now sleep-driving. Piecing her actions together, she realised that she must have gotten up, got dressed, unlocked the front door, driven the car for some distance, before reverse parallel parking into the same spot and getting back into bed. 'I have no idea where I went or how long I was away, but I came back and parked in exactly the same place,' Jackie explains. 'Nothing was out of order when I got up in the morning.' Anxious about the possibility of harming herself or others by driving through the streets of Seaford in her sleep, Jackie made an appointment with her doctor. Having told the story of her sleep-driving episode, the only thing her general practitioner could suggest was locking her up. 'They said, "Oh dear, we'll have to find you a secure unit,"' Jackie remembers. 'And I said, "Oh no, I'm not a criminal," and I'm afraid I got up and walked out.' Desperate for a solution, Jackie thought laterally. She had recently started a new relationship with a man named Ed, also in his seventies. One morning, Ed had woken to find the front door ajar, having locked the door before retiring to bed. It was apparent that Jackie's nocturnal wanderings were not waking Ed up in the night. In fact, when I talk to Ed about Jackie's sleep issues, he chuckles in a slightly bemused fashion. Apart from one or two events, when evidence of Jackie's sleepwalking is obvious in the mornings, Ed is largely oblivious at night, as, ironically, he sleeps so deeply.

She hit upon the idea of rigging up the front door to her apartment with a bell. 'A friend of mine very helpfully bought me a bell that I have on the chain lock. I hang it on there so that hopefully my partner, if I happen to go out of a night, will hear the bell jingling and will be able to stop me from going anywhere I shouldn't be.' In principle, this should have worked, but frustratingly Ed was such a heavy sleeper that he did not stir. A louder bell or buzzer was not an option for Jackie: 'The noise I'd need would be loud enough for the rest of the neighbours to hear and, you know, you can't upset them,' she confesses. After trial and error, Jackie and Ed have finally hit upon a solution for her sleep-driving. Jackie has invested in a safe, in which she locks away her front door key and house key. Her fear of opening the safe in the night has resulted in one other important adaptation: the safe has a time-lock. Once locked for the night, it will not open again until 6 a.m. – the ultimate safeguard for her. 'A neighbour down the hallway has spare keys for the car and for the door, because if anything happened in the night whereby we had to get out . . .'

It is difficult to know if Jackie is still getting up in the night but unable to get out of the apartment or if she is simply not getting up. Perhaps she is somehow subconsciously aware that she does not have the keys available to her. Ed's deep sleep means that he is not really a reliable witness to this.

Interestingly, Jackie says that she thinks the reason she stopped wandering around the cruise ship a few years earlier was because she somehow knew that if she went out of the cabin she would not be able to get back in; I had assumed in error that it was because she was locked in her cabin. Additionally, she tells me of a time when she left her flat in Seaford in the middle of the night and found the main door to the block wedged open in the morning. She assumes she did this in her sleep so that she could get back in. This certainly sounds like some degree of awareness, and perhaps supports the view that her brain is largely awake. If that is the case, you might expect her to wake feeling tired, as many sleepwalkers do, but instead she feels refreshed, none the wiser for these events.

* * *

So, these bizarre behaviours arising from very deep sleep have an underlying neurobiological explanation. The human brain has the ability, at least in some people, to be awake and asleep at the same time, with different areas of the brain in different states simultaneously. In fact, this situation probably arises much more commonly than we think. In very recent years, it has become apparent that this mixed state extends into the middle of the day. 'Local sleep', sleep activity only affecting small areas of the brain rather than the entirety of the brain, seems to occur in full wakefulness.

When kept awake for prolonged periods, rats demonstrate neurones in the cerebral cortex that fall silent for brief periods, in the same way as they do in deep sleep. These 'off-periods' are associated with increased errors during certain tasks. Similarly, when sleep-deprived, the human brain exhibits changes in activity suggesting very focal areas of sleep continuously occurring in the cortex, meaning that when we are tired, we really are 'half-asleep', or perhaps 'one-tenth asleep'. These changes, like in rats, are linked to performance errors when undertaking certain tasks, and would certainly explain why we feel less capable when sleep-deprived.

* * *

But why do some adults sleepwalk and others not? The short answer is that we just do not know. What we *are* aware of is that adults who sleepwalk often have other people in their family who also do it. A study of sleepwalking in twins has shown that if your twin sleepwalks, you are much more likely to sleepwalk if you are an identical rather than a non-identical twin. This of course implies that something in our genes predisposes us to sleepwalking and other non-REM parasomnias. In a single four-generation family containing twenty-two sleepwalkers, researchers found a large region of chromosome 22 to be linked to this condition. The region contains twenty-eight genes, and so far, efforts have not been able to identify a specific genetic cause. Other studies have shown an association between sleepwalking and other non-REM parasomnias, and an area of the genome encoding the HLA system, a grouping of genes implicated in immune function. Why this should be related to sleepwalking, though, remains a mystery.

But genetic predisposition is not the only explanation.

A common feature of people with non-REM parasomnias is the worsening by certain environmental or lifestyle factors. Many of my patients report that sleep deprivation, alcohol, daytime stress or anxiety can increase the likelihood of events. So, we know that these events are as a result of incomplete waking from deep sleep. In theory, therefore, there are two possibilities as to why these events might arise.

The first is that people with a genetic predisposition to sleepwalking might be partially woken by events that for a non-sleepwalker would have little bearing on their sleep. Their deep sleep may be inherently less deep or less stable. I have certainly seen patients in whom sleepwalking has been triggered by noises like a squeaking bed, a plane flying overhead or the rumble of a distant lorry. Perhaps a vibrating phone on the bedside cabinet when a text comes through in the middle of the night.

Daytime stress may cause night-time sleep to be less deep, too. Alcohol is generally considered a sedative, but actually may have precisely the reverse effect. It makes sleep more fragmented, and a full bladder can certainly wake people up – as can the snoring associated with a skinful of beer. Equally, all of us are aware of the effects of daytime stresses on sleep, making it more broken. Thus, in some people, anything that causes an arousal – a small shift out of deep sleep – may make a sleepwalking event more likely to occur.

The second possibility is almost the opposite explanation – that sleepwalkers sleep more soundly than non-sleepwalkers, and what would normally wake people up completely from deep sleep only succeeds in partially waking the brain of sleepwalkers, resulting in these events. Sleep deprivation is a potent way of making deep sleep even deeper, and some drugs commonly used as sleeping aids can certainly trigger sleepwalking events, sometimes even in people who have never sleepwalked. The most dramatic case of this I have seen is a woman in her seventies who began to sleep-bathe in the middle of the night (she was discovered chin-deep in bubble bath) after being prescribed a sleeping tablet.

The reality is that both explanations are probably valid. For children, in whom sleepwalking and other non-REM parasomnias are so common, deep sleep is incredibly stable, as any parent who has picked up their child in the middle of the night will testify, and this may well be why so many children exhibit these behaviours. In adults, the disruption of deep sleep by a variety of factors may well be a more significant explanation, on a background of an underlying genetic susceptibility. In fact, targeting these factors is often a successful way of treating non-REM parasomnias in adults. Cutting down on alcohol, reducing daytime stress, avoiding loud noises and an uncomfortable bed may help. Regular sleep patterns and avoidance of sleep deprivation may also be useful, as this reduces excessive deep sleep, from which people might have difficulty fully waking. Treatment of snoring is often a useful technique. For some people, however, particularly if you have types of behaviour that put yourself or others at risk, medication may be the only option.

So what have Jackie and Alex done to treat their parasomnias? To my surprise, Jackie has never really wanted any treatment for her sleep-driving. She wanted an explanation for why she does the things she does, but has worked out her own ways of keeping herself and the other road-users of Seaford out of danger. In many ways, this is perhaps one of the most convincing aspects of her sleepwalking for me.

Until recently, she has never really sought medical help for her events, but has always seen this as part and parcel of who she is. She has absolutely nothing to gain from this, and is somewhat taken aback by my interest in her condition. She tells few people about her nocturnal life, and has not really let it impact on her daytimes. Her primary concern is her and others' safety. Having been diagnosed with mild sleep apnoea as well, previous attempts at treatment have been focused on this aspect of her sleep, on the basis that occasionally stopping breathing might be precipitating her events.

She has been trialled on a treatment called CPAP, whereby a mask that is strapped to the face delivers pressurised air to prevent her airway collapsing. This treatment is intrusive and can be difficult to tolerate, as Jackie has discovered. She just didn't get on with it. An alternative would be an oral device, a little like a boxer's mouthguard, to push the lower jaw forward and create more space at the back of the throat. For the moment, Jackie has not pursued this, nor is she keen to take any medication for her sleepwalking. From her perspective, using her time-locked safe has resolved the issues that cause her concern.

For the most part, Alex finds his sleepwalking episodes amusing and is generally relaxed about the sleep terrors. In recent months, his nights have been a little more peaceful, for some unknown reason. Perhaps he is one of those individuals who is slowly 'growing out' of his non-REM parasomnias, albeit a little later on in life than most. He is rather phlegmatic about his sleep issues. 'I've had it all my life, so it's one of those things I've just lived with,' Alex says. 'I don't know anything different. Whenever there's a party and someone has to say something weird about themselves, then I usually have one of those stories.'

Alex is keen to try non-drug-based treatments in the first instance. One of his acquaintances has suggested hypnotherapy, which Alex is going to pursue. We have agreed that if this does not help, the next steps would be to start medication. Initially, I have suggested melatonin, a tablet version of the hormone produced by the brain as a signal to sleep – what we term a sleep promoter. In many countries around the world, this drug is seen as a health supplement, and can be bought off the shelf in supermarkets and pharmacies, although in the UK it is only available on prescription.

If he is still putting himself at risk, then other options would include antidepressant drugs or benzodiazepines, neither of which are without issues. The decision to use these drugs needs to be weighed up carefully. We do not fully understand how these drugs work, and the evidence for drug treatment in non-REM parasomnias is for the most part limited, due to lack of research. Any treatment needs to be supplemented with avoidance of potential triggers such as noise and sleep deprivation, and practical measures such as door alarms and locks on the windows.

When pressed as to why he thinks he might be having sleep terrors, he says: 'A lot of people have given me their theories on it – they think that I've been bottling up stress and anxiety and then letting it out when I'm asleep – but I'm not really sure.'

To my mind, Alex is one of the most relaxed people I have met in my clinic, and seems largely unruffled by most things. Katie, his ex-girlfriend, confirms my view: 'I don't think I've ever seen him anxious about anything. In the six years I've known him, I've never known him really to repress anything. He's always been quite open about any issues that he's having.'

'I guess it helps me with stressful situations in real life,' Alex muses. 'I know that if one day I do wake up and there is a man in the room with a knife, I'll be more relaxed about it and hopefully have a clearer head and deal with the situation. So I guess it's sort of preparing me for that . . .'

* * *

Our previous binary view of brain states, either awake or asleep, really does not represent the truth. Rather than black and white, it appears that there are infinite shades of grey – and Jackie, Alex and many others like them occupy this grey zone in the middle of the night.

3

DISNEY WAS RIGHT

In my childhood, I must have seen the Disney film *Cinderella* several times. I have vague memories of a scene in which Cinderella is dancing, and her animal companions help her to make a dress. Birds clutch a ribbon between them as the mice busily stitch it together. Some thirty years later, however, watching the film with my two young daughters, I was suddenly struck by something in the film.

If you have exceptional recall, or have young children too, you may remember two scenes. In one, Bruno the dog is sleeping on a rug on the stone floor of the kitchen. As he does so, he growls and his legs move, as if he is running or chasing something. He is dreaming of chasing his arch-enemy Lucifer, the stepmother's evil cat, and he suddenly leaps up, bites the rug he is lying on, before waking up, startled. Cinderella pets his head as he calms down. In another scene, the King, obsessed with marrying off his son, Prince Charming, lies asleep in the centre of his vast bed, which has an ornate baroque golden headboard and royal red bedcovers. Lying on his side, with his head on the pillow, he laughs and chuckles as he dreams that his imaginary grandchildren are riding on his back. His dream-grandson, clutching a golden sceptre or rattle, begins to bang him repeatedly on the head. The King wakes with a start and tumbles out of bed, caught up in the bedcovers.

* * *

A couple of years ago, I had received an email from a psychiatrist colleague about a patient. 'He is eighty years old. The main issue is what he calls "violent night terrors", occurring between 2 and 4 a.m. over the past six months. He sleeps with his wife who he says has osteoporosis, and has been so concerned he has been thinking of moving into a separate bedroom. There is a background of some mild "difficulty with sleep" of a longer duration than six months.'

When I first meet John* and his wife Liz*, it is quite clear they are both traumatised by their recent experiences. John is in his eighties, but looks much younger. I envy him his full head of grey hair. He is still actively working and is incredibly successful in his professional life. He is tall and slim and wears fashionable round glasses with thick, dark-brown frames. He tells me his story in a soft, deliberate manner, despite the obvious angst this causes him. Liz sits next to him, elegant and equally gentle, trying to communicate her concern and the shock of what has been going on, mindful of upsetting John. She explains the final straw that caused them to seek medical help.

'This time he'd actually come all the way over to my side of the bed and grabbed me on my arm, and it scared me so much when the nails went in. I was out of bed before I knew it, screaming my head off – and I'm not a screamer. And I was shaking and crying, just saying, "We can't do this any more." That was the final one, when I realised it was too much.'

John had managed to draw blood. It was at this point that they had moved into separate beds.

But this episode was not the first. John has been having these 'night terrors' for a couple of years now, rather than for the six months mentioned in the referral letter. 'I think there has been a leading-up to this for years,' Liz explains. 'Before the actual physical violence, there would be this very weird sound coming from John, almost from the depths.' Over the months, John's episodes intensified.

Liz tells me: 'Occasionally, he would kick me, but it would be with such ferocity that it would be as if you were walking behind a horse or a donkey and they lashed out and kicked you. And it would be so unlike John. That was the freaky thing, because it was almost like the day person was completely different to the night person, so that added to the weirdness of it all.'

John shifts uncomfortably and grimaces on hearing what he has put his wife through. Liz continues, talking about the strange sounds John would make: 'And it would be like a crescendo, and it was always scary. What was he going to do once he'd got to the top, you know, with the crescendo? Was he going to lash out or what was going to happen?'

These events were happening every week or two. Even after moving into separate beds, Liz could still hear John thrashing about in bed.

John tells me: 'Although I knew that I was having dreams about things, I didn't realise that they were manifesting themselves in such obvious ways – thrashing about, leaping out of bed, calling out in a very strange voice and all of that kind of thing – until Liz pointed it out.'

I ask him about the content of his dreams.

> It's the kind of traditional nightmare, like suddenly finding yourself in a large wood with a tiger. What happens is that I'm doing something perfectly routine, something perfectly normal – there was no reason I shouldn't be in a wood, and I was often in a wood, you know – but then

> suddenly you're aware of a presence that you weren't expecting and then it becomes sinister and you get fearful. It could be other animals, or the sort of things that frighten people like snakes and anything like that. Things that eat you and bite you. The typical experience is that at that moment when you expect their jaws to close on you, that's when you wake up.

According to Liz, John sometimes lets out what she describes as a disembodied howl as he lashes or kicks out. 'I'm kicking at some object, usually an animal, to get something away.'

Liz's initial fear was that John was suffering from epilepsy, and John really didn't know what to think. 'I'm always embarrassed by them now. Since my wife started to point out that I was having these things and being violent in the process, I've rather dreaded them.'

On listening to their story, however, it is quite clear to me that they are describing John acting out his vivid dreams or nightmares. He experiences dreams of a narrative structure, a story evolving like the plot of a film or book, which suddenly turns nasty. The fear or strong emotion triggers the kicking or lashing out – a natural response to what is going on in John's mind.

Their first thought had been to seek psychological help. John explains: 'At one stage I went to a therapist that Liz recommended through a friend of hers.' The therapist was obviously taking the Freudian view that dreams are a window into the subconscious, and perhaps represent repressed thoughts or psychological trauma. 'I had several sessions with him [the therapist], which was all about going into tigers and what tigers meant and stuff, a great deal of detail. But it didn't do anything for me, I'm afraid.'

* * *

If the psychological explanation for John's nocturnal problems is insufficient, could there be a neurological explanation? As discussed in the previous chapter, most of our dreaming, particularly these dreams with narrative content, arise in a stage of sleep termed REM sleep. The discovery of this stage of sleep was, as is often the case, rather serendipitous.

Eugene Aserinsky, one of the discoverers of REM sleep, recalls meeting Nathaniel Kleitman – at the time already one of the most prominent sleep researchers in the world – in his small office at the University of Chicago in the early 1950s. Aserinsky was a graduate student searching for a research project. He self-effacingly recalls:

> After satisfying himself that my mind was a clean slate devoid of any self-generated ideas, Kleitman related a story. He had read in *Nature* [one of the most highly regarded scientific journals to this day] an article by a physicist called Lawson, who claimed that, while riding in a railroad compartment, he was able to distinguish sleep onset through observing the blinking rates of fellow passengers. What disturbed Kleitman was the allegation by Lawson that the blinking stopped abruptly with sleep onset rather than stopping gradually.

Aserinsky was more disturbed by the fact that *Nature* had bothered to even publish such an un-noteworthy observation.

Nevertheless, he was tasked with reading all that had been published on the subject of blinking – thus becoming 'the premier savant in that narrow field' – and testing Lawson's hypothesis. Over several weeks, Aserinsky tried fruitlessly to build a machine to record eyelid movements, eventually admitting defeat. Instead, Kleitman suggested that the young researcher spend some time observing babies as they slept. (Kleitman was already involved in a study involving recording body movements in sleep in infants, using kit that was attached to their cribs.) A short while later, Aserinsky had to return to his mentor, his tail between his legs, admitting defeat yet again. He had noted that babies' eyelids quivered in their sleep even with their eyes closed. Did this represent a true blink?

Aserinsky writes with a lovely turn of phrase: 'I plodded along for months attempting to draw blood from this research turnip . . . but painstaking, diligent exploration of minutiae will frequently lead to the "golden manure" phenomenon whereby there is a rewarding result.' After several months of observations, he noted that roughly every hour or so, infants had a period of about twenty minutes during which their eyes would stop moving. It was some time later, having had help in building an eye movement recording device, that he began to successfully measure eye movements in sleeping adults.

He started to witness vigorous jerky eye movements in subjects who appeared to be deep asleep, and that these periods were also episodic throughout the night. On one occasion, he recorded a subject who appeared to be having a nightmare during one of these stages of sleep characterised by these eye movements, with him moaning or talking in a slurred way. On waking the subject, and hearing the description of his nightmare, Aserinsky assumed the speech and his nightmare were clearly related, and began to wonder if the eye movements related to visual imagery while dreaming.

Over the next few years, he went on to perform countless experiments, some in conjunction with William Dement, one of the other grandfathers of sleep research, including at one point a televised sleep session using his son as the subject, and found that subjects would frequently report dreaming if woken from REM sleep.

Aserinsky moved away from the field of sleep research for about a decade, but his interest was reignited when he witnessed his St Bernard dog, coincidentally, just like Cinderella's companion, also called Bruno, going into what was by this stage called REM sleep, exhibiting regular muscle twitching. He assumed that Bruno was repeatedly dreaming the same dream, and that there was some association between his muscle activity and the content of Bruno's dreams. And so Aserinsky's research into REM sleep was driven on by his dog doing much the same thing as John.

* * *

Listening to John recount his dreams, I know they are typical of those arising from REM sleep. But the type of dreams he is having will sound familiar to everyone, so why don't we all act out our dreams? REM sleep has subsequently been the subject of a huge amount of research, but some sixty-five years later, the function of REM sleep remains a mystery (see Chapter 13). However, we do know that during this stage of sleep, which occurs about four or five times per night, our brainwaves look like we are almost awake, and there are changes to how our blood pressure, heart rate and breathing are regulated. But, despite our brains appearing to be so active during REM sleep, our bodies are largely paralysed. All muscles, with the exception of the muscles of the eyes, the diaphragm (the large muscle that is responsible for most breathing movements) and sphincters at the top and bottom of the gastrointestinal tract, develop complete weakness. The muscles essentially shut down. This is such a strong feature of REM sleep that in the practical setting of the sleep lab, we use the drop-out of electrical signals from the jaw or limbs to help define it.

So what is going wrong with John's sleep? It is clear that John's lashing out or howling in the night is a manifestation of dreaming, the type of dream associated with REM sleep, but why is the paralysis not kicking in, leaving Liz at risk of a kick or punch when he is dreaming of tigers, snakes or arguments?

John and Liz give a classical description of REM sleep behaviour disorder, or RBD for short, where that mechanism of paralysis goes awry. RBD is thought to affect less than 1 per cent of the population, and typically comes on in the fifties or later. The disorder is a bit of a curiosity, as it is one of the few medical conditions described in animals before humans.

In the 1950s, Michel Jouvet, at the time a researcher at the University of Lyon studying sleep in cats, began to undertake experiments to better understand how REM sleep is controlled by the brain. During the course of his experiments, he found that damage to a part of the brain called the brainstem, the slender tract that is the conduit of all brain signals to the rest of the body, caused cats in REM sleep to hiss, walk, fight or behave as if they were chasing prey. They continued to show brain activity consistent with REM sleep, but the muscle paralysis expected simply did not materialise. In fact, these cats behaved exactly like Bruno, the dog in *Cinderella*, chasing Lucifer the cat.

Amazingly, *Cinderella* had been released in 1950, some fifteen years before Jouvet published his results. It seems that Disney, or at least some of his writers, were scientific visionaries, or at least keen observers of nature.

Researchers speculated that if this sort of thing could happen in cats (and cartoon dogs), then in all likelihood it could occur in humans. Indeed, this kind of behaviour was beginning to be described in scanty case reports in the medical literature in the 1960s, but it was not until 1986 that this condition was formally defined and accepted as a disorder, named REM sleep behaviour disorder.

In some respects, RBD sounds a little like night terrors or sleepwalking. But witnessing it, as we sometimes do in the sleep laboratory, shows it to be quite different. During RBD, the eyes are closed, and the movements often appear purposeless, simple flailing or lashing out. There is little interaction with the external world, and people do not get out of bed. Speech is often not fully formed, usually largely unintelligible, and may be in the form of shouting or swearing. Compare this to sleep terrors or sleepwalking, when people often get out of bed and their eyes will be open as they pick up objects and sometimes act out complex tasks, like the driving we saw in Chapter 2. Speech in these sorts of conditions is also different, in that it is usually intelligible, with fully formed sentences. Sometimes people can have limited conversations, albeit ones that are often nonsensical.

The nature of the dreaming is also different. In RBD, the dreams are stories developing over time, while in sleepwalking or sleep terrors, it is often visual imagery, disjointed in nature – walls crashing down, natural disasters, insects, spiders. Another feature that differs is the timing of these events. Most of our deep sleep is in the first half of the night, and most of our REM sleep is in the latter half. So, while sleepwalking and night terrors often occur in the first few hours of sleep, RBD is much more likely to happen later on, in the few hours before waking.

* * *

John has a range of dreams or nightmares that culminate in

RBD. 'I had one a month or so ago which did involve being extremely angry with somebody who I'm quite frequently angry with. That was a reference to a real-life person,' he tells me. 'And at the point I leapt out of bed, I'm sure I was on the point of punching him on the nose. For a pretty non-violent sort of person, that was rather uncharacteristic. It was a gesture of complete frustration – of not being able to understand something, not being understood, not being able to get through to somebody. No doubt if I'd continued with the dream I would have landed a punch on somebody.'

Other dreams are more along the lines of being attacked by animals, like the tiger he has previously told me about. 'I'm quite fearful of water. One of my horrors is the idea of swimming off Bondi Beach and being chased by a shark or something like that. I'm sure that's cropped up. I'm obviously rather cowardly about big animals, and big animals have occurred more than once in my dreams.'

Much like Jouvet's cats, the behaviour that John exhibits includes fighting or running away. But, unlike the cats, there is nothing to suggest brain damage in John. He is active, physically and mentally, without a hint of any brain disorder.

We used to term RBD as 'idiopathic', meaning without obvious cause. In some patients, this is thought to have been triggered by certain medications, predominantly antidepressants. We occasionally see families with multiple members affected by RBD, suggesting that at least sometimes there may be a genetic component. More recently, we have also learned that in young people it can be a feature of narcolepsy, a neurological disorder that causes loss of control over sleep and dreaming. It is very rare to see it as a result of obvious damage to the brainstem, although I recently saw a young man infected with HIV since birth, with a tumour of the midbrain and brainstem, who vigorously acted out his dreams throughout the night. Perhaps the commonest cause of RBD is when it disrupts the sleep of patients with known degenerative brain disorders such as Parkinson's disease or a related condition called Lewy body dementia, where features of Parkinson's, such as slowness of movement and tremor, co-exist with cognitive decline and hallucinations.

In the past few years, however, our concept of RBD has been revolutionised. Rather than the majority of people having 'idiopathic' RBD, we are learning that it is often a *precursor* to these neurodegenerative conditions; a very early warning sign of brain conditions like Parkinson's disease and Lewy body disease. Why this should be the case is slowly becoming apparent.

The brains of people with these disorders, when studied under a microscope, show deposits of a protein called alpha-synuclein. When this protein is deposited in the basal ganglia, an area deep in the brain fundamental to the control of movement, it causes the overt features of Parkinson's disease that are most familiar to us, like tremor, slow movement and walking difficulties. But studies of brains of people with Parkinson's have shown these protein deposits elsewhere in the nervous system. These include the nerve that is responsible for smell, the olfactory bulb, and the nerves that supply the gut and blood vessels. Of most relevance, these deposits are also found in the brainstem, the area responsible for switching on the paralysis seen in REM sleep. And, in fact, these changes are seen in people years or even decades before the onset of overt Parkinson's or other degenerative conditions of the brain. So it makes perfect sense that some people will develop RBD in the very early stages of these brain disorders.

In keeping with the locations of deposition of alpha-synuclein in the nervous system, other symptoms that people frequently complain of years before developing Parkinson's disease include loss of sense of smell, severe constipation, bladder problems and blood pressure issues. Now, of course, not everyone who has RBD goes on to develop Parkinson's or other brain disorders, and loss of smell and constipation are incredibly common symptoms, but if we can use these conditions to predict a high risk of developing Parkinson's disease, then this gives us a window of opportunity. While at present there are no drugs that can prevent or slow the progression of these brain conditions, there are ongoing trials. And it makes sense that, if these drugs become available, candidates for these medications would be those at high risk of developing these disorders, before sufficient damage has been done to the brain to cause more overt features. So, in the future, if I see a patient in my neurology clinic who is of a certain age, with constipation, loss of smell, RBD and other problems, I may simply plug his or her details into a computer which calculates the risk of developing Parkinson's in the next five or ten years, and decide whether or not to start a drug to prevent this. Efforts are currently ongoing to develop these types of algorithms, to clarify the risk of developing Parkinson's for any particular individual, but RBD, and other problems, may be fundamental to us treating these disorders in the coming years.

But what about John? He has absolutely no features of Parkinson's disease or any other degenerative brain condition. He lost his sense of smell a number of years ago, and has some bladder problems, although this is likely related to a previous bladder operation.

* * *

John's anxiety about his wife's safety is understandable. Liz is troubled by osteoporosis, brittle bones, meaning she is at risk of fractures. He is worried that he will cause her a serious injury. 'I've been woken up by the prospect of injuring somebody else, but I haven't injured myself,' he says. 'I obviously take it so seriously that until I can find a solution to this, we're going to have to continue sleeping in separate beds. I've been sleeping on the sofa bed for the last nine months. And I'm not intending to go and subject her to a sleepless night until I know that this is not going to happen again.'

Additionally, Liz's fear of what might happen at night means she finds it difficult to sleep in a bed with him. 'It was just very frightening. I felt it wasn't safe to actually fall asleep,' Liz explains. 'Because we now sleep in separate beds, I've noticed I'm sleeping much better, because obviously I was more fearful than I was aware of.' On a number of occasions, before this separation, she had woken up in the morning with bruises and scratches.

The cessation of sleeping in the same bed, while keeping Liz safe and comforting John, is clearly not entirely welcome. It must be a huge psychological blow to any couple not to share a bed, and I suspect this change is tinged with sadness for them both. Despite this, Liz does find glimmers of humour in the situation. 'The only one time I did giggle was, he was lying on his back, and I was relieved he wasn't lying on his side and facing me, and he kicked out and fell off the bed. But apart from that, it hasn't been at all funny.'

* * *

There are two things that particularly strike me about John and many other patients with RBD. The first is the nature of the dreaming. John and others clearly describe a dramatic change in the content of their dreams with the onset of this condition.

These dreams or nightmares of being attacked, fighting or running away are fairly consistent. Occasionally we do see other sorts of activity, like acting out smoking a cigarette, petting a dog, laughing or singing, but these are much rarer. The behaviours that we see are the very characteristic movements of thrashing, flailing or kicking that Liz describes, often accompanied by swearing, shouting or screaming. So what explains this striking alteration in dream content? It certainly does not appear to be a function of daytime aggression. Liz says: 'John never really expresses anger. I very rarely see him angry, so that's also a very strange thing because at night he is very angry and very frightened.'

I perceive John to be a very gentle and mild-mannered man, and indeed multiple studies have shown patients with RBD to have very low levels of aggression in their waking lives. Therefore, could the changes in the brain that occur in RBD directly be responsible for the change in dream narrative? There is some evidence hinting that this might be the case.

A research study has shown that patients with Parkinson's and RBD are less aggressive during the day than those with Parkinson's and no RBD, at least suggesting that changes within the brainstem have some role in aggressive tendencies, but certainly this does not give us a full explanation. An alternative, relatively new, hypothesis is that the changes within the brainstem give rise to these flinging or thrashing movements. In the same way that we can integrate sensory stimuli into our dreams – a clap of thunder in the outside world becoming an explosion in our dreams, or a touch on the leg morphing into an animal clawing at us – perhaps these movements, these violent actions, may influence the stories unfolding in our REM sleep.

RBD is also thought to be more common in men than

women. Indeed, men make up about 80 per cent of patients referred to sleep centres. Is this because of the effects of testosterone on our brains? Or is RBD simply milder in women than in men? It has definitely been shown that unpleasant dream content in men is more likely to involve verbal or physical self-defence, while in women it is more likely to be a case of a sensation of threat or fear, of running away. So could this mean that women are less likely to lash out, and less likely to injure themselves or their partners? Or could it simply be that women are more embarrassed to seek medical help, or more likely to push their partners to seek attention? We just don't know.

The other remarkable feature is the contrast in movement between day and night. As Liz comments of John's episodes: 'It's almost unbelievable that he has that sort of force in him. If you asked him to do that in daily life, he wouldn't have a quarter of the strength.' In his case, I suspect this is simply a function of the level of stimulation, the fear of being bitten by a tiger or attacked by someone, that releases levels of activity we rarely need to rely on in our daily lives. I think of stories of people exhibiting enormous feats of strength when under intense pressure, like lifting a car to free an injured pedestrian. But what is truly amazing is that people with Parkinson's disease or related disorders may be crippled by their condition in the day, but may move fluidly and quickly during their RBD. The tremor, slow movements and quiet voice during the day are replaced by vigorous, quick actions and loud shouts at night. It is almost as if their Parkinson's disease is transiently cured in REM sleep!

In one of the first studies of this phenomenon, the researchers described one patient who was witnessed squatting on the bed, waving his arms as if flying, emitting a siren noise with a duck's voice. The patient, who during the day had been unable to squat down, was very slow of movement and had a quietened voice, had been dreaming of being a police-duck, flying after a pigeon-thief. Other patients with significant daytime symptoms related to their Parkinson's have been witnessed engaging in sword fights with an invisible foil, furiously rowing an imaginary paddle to escape from crocodiles in a canoe, or delivering political speeches in a loud voice. But how is this possible?

Some patients may exhibit a phenomenon called 'paradoxical kinesis', where movements suddenly normalise under life-threatening circumstances, for example a bedridden patient who suddenly manages to get up and flee his house during a fire. So perhaps in some cases with RBD, the strong emotional content of their dreams may override their Parkinson's disease. But clearly this would not explain a flying police-duck chasing after a thief, or other examples, such as singing a pop song in the middle of the night while dreaming of being in the shower. It remains speculative, but some researchers have suggested that, in REM sleep, signals from the motor cortex, the part of the brain where movements are generated, simply bypass the basal ganglia, the areas involved in Parkinson's disease that regulate movement in waking life. Why this should be the case is another of the many unknowns in neuroscience, but remains a fascinating question.

* * *

In addition to John's RBD, his sleep study has demonstrated mild sleep apnoea – snoring to an extent that he intermittently partially obstructs his airway – and this, as well as getting up at night to pass urine, a sequel to his prostate operation, contributes to his poor sleep. For his RBD, however, he has now started on medication. The mainstay of treatment used to be a sedative called clonazepam, a benzodiazepine similar to Valium, which for some people works very well. It can have side effects, however, and can leave people feeling groggy and lethargic in the morning. For people with neurological disorders like Parkinson's or Lewy body disease, it can cause confusion or predispose them to falls.

In recent years, we have started to use melatonin in the first instance. This gentle drug, as with sleepwalking, seems to work for a significant proportion of people with RBD, with many fewer side effects. Numerous potential ways in which melatonin may work have been proposed, but its mechanism of action remains uncertain. John has been on melatonin now for a number of months, with some benefit.

John tells me: 'They're reducing in frequency since I started taking melatonin. And in fact they're pretty infrequent now.' Liz agrees: 'Since taking the medication, I'm daring to actually believe maybe it's doing something, maybe he is sleeping a little bit deeper, but not so deep he doesn't know what he's doing. I did hear him crying very loudly, like a very strange cry, a couple of weeks ago in his sleep, and I've also in the same week heard him laughing his head off in his sleep. So, you know, it's not quite as violent, but it's still something that's coming out.'

John and Liz are still not confident enough to share a bed, but progress is being made. We are slowly increasing the melatonin dose, and trying to treat his sleep apnoea and urinary symptoms. They live in hope that they will get to the point where they can sleep in the same bed. However, giving John a clear diagnosis has also had other benefits. The shame and embarrassment that John felt when he first started having these episodes is waning. Liz sums it up succinctly:

There's quite a lot of shame involved in this. From the person actually doing the kicking, beating up the person, your partner, your wife, your husband, whoever it is who's got it. My husband would say, 'No, don't tell anybody. Don't tell anybody about this!' And I just had to say, 'I'm not going to keep it a secret. Because it's not something that one needs to be ashamed of. It's something that you're unconscious of. It happens and you don't do it on purpose.'

* * *

It seems, therefore, that Disney was right. Bruno the cartoon dog, as well as the King, exhibited RBD years before neuroscientists even reported it in cats, and decades before it was formally recognised by medics. What could be seen as artistic licence turned out to be seminal medical documentation. And, watching *Cinderella* all those years later, I wondered how many people might have seen this with their children or grandchildren and recognised something of what they or their partners do in the night.

Postscript: In the past few months, John and Liz have resumed sharing a bed, facilitated by a large bolster dividing them. Liz says that there have only been a few episodes; John on one occasion roaring like a lion, but without kicking.

4
RUMBLINGS

When I look through the medical records of my patients, they are full of abbreviations or acronyms. Some, like CADASIL (cerebral autosomal dominant arteriopathy with subcortical infarcts and leukoencephalopathy), or SOREMPs (sleep onset rapid eye movement sleep periods), are absolutely necessary, to save ink, time and the sanity of the poor medical secretary typing clinic letters. Others, much beloved by medical students and unspeakably rude or insulting, these days largely only exist as urban legend, having fallen out of usage decades ago. And then there are those that describe what the patient is telling you.

Of all the acronyms in common use, none is more guaranteed to sink the heart of the doctor opening the notes than TATT – 'tired all the time'. When the presenting complaint is TATT, you know that the consultation is going to be a long one. The list of possible causes of TATT is almost endless. Underactive thyroid gland, diabetes, depression, chronic fatigue, cancer, shift work, anaemia, autoimmune disease, carbon monoxide poisoning . . . it goes on and on. Some of these causes are life-threatening, while others are minor complaints, but TATT is incredibly common, so for the general practitioner, an inordinate amount of time is spent filtering in order to find those with a serious underlying problem: a needle in a haystack.

I am sure that these four letters – TATT – were scrawled in Maria's* medical notes when she first went to visit her doctor, whose spirits undoubtedly ebbed as he or she listened to her story. When I meet Maria, she is eloquent, vivacious and energetic. But the picture she paints of only a few months ago is hard to envisage.

She tells me: 'So, over the last couple of years, I was just exhausted all the time. I had no energy. I didn't feel well and I'd been going to the doctor. I thought maybe it was my thyroid, or maybe I had anaemia, and so I went repeatedly for blood tests and things. I'm overweight and was having real difficulty in losing any weight no matter what I was doing. I had no energy to exercise.'

Maria, now in her forties, was juggling motherhood and a challenging job in healthcare, but her sheer exhaustion was impacting all aspects of her life. The sense of fatigue was all-pervading. 'I think one of the things that I found very difficult was that I veered between extremes of being very manic and just really furious. I had no sense of perspective. I had no resilience. I couldn't cope with anything.'

She could just about cope with work and home life, but any slight deviation, any additional pressure, was intolerable. 'Any little bump in the road was like the worst thing that had happened to anybody ever. I'm quite a cheery person by nature and my sense of humour went.'

Maria's lack of energy was having a significant effect on her relationship with her husband. She recalls:

> He's a very laid-back person, but there were a few times he said to me, 'This has got to change. This can't carry on.' But he knew I was doing everything that I could. I was trying meditation, I was doing mindfulness, I was trying

> to go for yoga classes, I was trying to change my diet. I even changed my job. There were lots of things that I was doing all the time to try to get back to who I should be.

As I listen, I am struck by the lengths that Maria went to in an attempt to improve matters. Changing your job is not a minor change, after all, but a real upheaval.

Her poor health was having physical and psychological consequences, too. Maria recalls her distress: 'My skin wasn't great. I'd always had really clear skin and then all of a sudden I developed eczema on my hands, also across my face and my eyes. I found that very upsetting. I did feel like I was falling apart. I thought, *That's it. I'm kind of freewheeling down to menopause and I'm just crumbling*. It felt very unfair.'

And for Maria's husband, it was a struggle too.

> My husband was worried, but he was also frustrated because he was having to deal with the fallout of me not being rational, not having energy to do things, and making plans and then not being able to follow up with them because I was too tired. It killed our social life because I just wasn't able to go anywhere. I was just too tired. So I'd make plans full of optimism and then, when the time would come, I wouldn't be able to go. You can only do that so many times before people stop making plans with you because you're a flake.

She recalls that, if she had a day off, she would take her children to school, come home and get back into bed, setting her alarm for twenty minutes before she had to pick them up again. 'I didn't have the energy to read a book because as soon as I started reading I'd be asleep before I got to the end of the page.'

Predictably, the problems weren't just confined to her home and social life. At work, Maria would have to pinch herself under the table in meetings to make sure that she didn't doze off. 'I would find it very, very difficult to stay awake, so I'd have lots of coffee, lots of glucose energy sweets to keep me going.'

Maria battled on, but repeated trips to her doctor did not result in any clear diagnosis, or any prospect of treatment. However, some eighteen months after first seeking medical help, Maria had a lightbulb moment. She remembers it vividly: 'I was at the doctor's for something else, and an information video came on the screen in the waiting room describing some symptoms, and I thought, *Gosh, that sounds like what I have!* So, when I went in to see my GP, I said, "I think I've got sleep apnoea!"'

* * *

Snoring is both the subject of comedy and an instrument of torture, particularly for those sharing a bed with a snorer. It is incredibly common. For most people, a narrow airway or partially blocked nasal passages disturb the airflow at the back of our mouths, causing the soft tissues in our throat – the soft palate, tonsils, adenoids and uvula (the dangly bit that hangs down at the back of the throat) to reverberate. As we breathe in, the turbulent airflow causes these areas of the airway to vibrate, resulting in anything from a gentle purr to volumes as loud as a passing lorry.

Obstructive sleep apnoea, however, is of a different order of magnitude, not in volume but in effect. As you drift off to sleep, the countless small muscles that contribute to the stiffness of the walls of your airway slacken a little. If your airway is narrow enough or becomes sufficiently floppy as you drift into sleep, it can actually partially or completely obstruct.

This collapse of the airway results in oxygen levels falling and the heart rate increasing, but from a sleep perspective causes a disruption of your sleep. As the depth of your sleep is fragmented, muscle tone in the airway returns briefly to allow you to breathe again, and this cycle can continue throughout the night. These obstructive events can happen ten, twenty and, in rare cases, even 100 times per hour. And if your sleep is disrupted at these sorts of levels, it is entirely understandable why you might wake up tired.

Maria says, 'I knew that I had an issue with snoring. My husband had said that my snoring was particularly loud, but I dismissed it because, in my experience, lots of people snore, so I didn't think it was that big of a deal. A bit of a comedy thing, snoring, so I didn't treat it with probably the gravitas that I should have. I didn't take it seriously.'

Nevertheless, the evidence of Maria's sleep disruption was in plain sight. Her snoring was loud enough for her husband to regularly seek sanctuary in their toddler's bedroom, and he had commented that occasionally she would stop breathing in her sleep.

While it was obvious to her that she had issues with her sleep, Maria had an alternative explanation.

> Although I was exhausted, I was sleeping. I'd been asked by my doctor, 'Do you sleep all right?', and I do. As soon as my head hit the pillow, that was it, I was out like a light – whatever time I went to bed. And although I used to wake frequently in the night, I thought it was a bad habit that I had got into from when my children were still breastfeeding. Neither of them slept through the night very well, so I thought I'd just got into the habit of waking up frequently. So I put it down to that. But I never felt refreshed, never.

The severity of sleepiness associated with sleep apnoea can be extreme. At our centre, we have had countless patients whose sleep apnoea has come to light after they have fallen asleep at the wheel of their car and had an accident. I vividly recall one patient who I saw in my epilepsy clinic. He had collapsed several times while standing, and was being investigated for seizures. On one occasion, he was standing next to his desk at work and remembered feeling a little sleepy. When he regained consciousness, he was lying on the ground, blood spattered across his desk and his face, having broken his nose as he slumped face first onto the corner of the table. Investigations for epilepsy came to nothing, but he was found to have terribly severe sleep apnoea. As soon as this was treated, his collapses completely stopped.

We are living through a sleep apnoea epidemic. A recent Swiss community study suggests that up to one in two men and one in four women have significant problems with breathing in their sleep. The rates of sleep apnoea have increased in parallel with our girths and neck circumferences. As we get larger and heavier, sleep apnoea becomes more and more common. In the UK, 62 per cent of adults in 2014 were deemed as overweight or obese, compared to 53 per cent a couple of decades earlier. This increase in size in recent years has been mirrored throughout most of the developed world. In the US, graphs of obesity in the US population show a relentless slope upward, particularly since the early 1980s.

This association between obstructive sleep apnoea and obesity has long been understood. Charles Dickens, that keen observer of medical Victorian London, wrote in *The Pickwick Papers* of the obese boy Joe: '"Sleep!" said the old gentleman, "he's always asleep. Goes on errands fast asleep, and snores as he waits at table."' It was this character who led to a variant of sleep apnoea known as obesity-hypoventilation syndrome to be called Pickwickian syndrome. Actually, there are even earlier references to sleep apnoea, or something like it, associated with obesity. Take Dionysius, the tyrant of Heraclea Pontica who reigned in the fourth century BC. He was said to be enormous and terribly somnolent, and his servants resorted to long needles to jab him awake. Another historical character, Magas of Cyrene, died in 250 BC, 'weighted down with monstrous masses of flesh in his last days; in fact he choked himself to death', according to rhetorician Athenaeus.

Increased weight may precipitate or worsen sleep apnoea in several ways. Local deposition of fat in the neck makes the airway narrower and more likely to collapse, but fat on the chest increases the effort of breathing, reduces lung volume due to weight on the chest, and increases the metabolic demands on the body. And weight loss often improves sleep apnoea.

But obesity is not the only cause. Sleep physicians will sometimes see people of normal weight with astounding levels of apnoeas – pauses in breathing overnight. While Maria is overweight, she is not obese. There are multiple other factors. Sleep apnoea may run in families, and is often related to the shape of the airway. A large tongue base or a recessed lower jaw will cause a narrowing, as sometimes do very large tonsils. Sleep apnoea appears to be more common in people of Southeast Asian ancestry, perhaps related to the shape of the head and resultant shape of the airway.

* * *

The effects of having your sleep disrupted several times an hour obviously result in excessive sleepiness and feeling unrefreshed during the day. This in itself can be dangerous to your health. Having sleep apnoea increases the risk of a road traffic accident by two or three times. But it is not just sleepiness that we need to worry about.

We know of many physical and mental consequences of sleep deprivation, and being partially woken throughout the night is simply another form of this. But these repeated suffocations also have costs above and beyond simply damaging your sleep. With each obstruction, there is a surge of noradrenaline, a rise in heart rate, blood pressure, and an increase of pressure in the chest, a stiffening of the arteries, and a drop in oxygen levels. We are slowly starting to fully appreciate the wider significance of these physiological changes happening hundreds of times per night. With each pause in breathing, changes to blood flow back to the heart result in alterations to a hormone called ANP, which is secreted by the cardiac atrium. The effect is that the kidneys continue to produce more urine than they otherwise would overnight, which results in more frequent nocturnal urination. Maria recalls: 'I'd get up probably three, four or five times in the night [to urinate].'

It has now been recognised that sleep apnoea is strongly associated with high blood pressure and the serious problems that it can cause, such as heart disease and strokes. Recurrent jolts of sleep disruption in themselves cause spikes in blood pressure, entirely independent of drops in oxygen, but it seems that intermittent low oxygen levels have an additive effect. In combination, the sleep disruption and effects on oxygen levels act to prime the body to generate high blood pressure during the day through changes to the sympathetic nervous system (the neurological mechanisms that utilise adrenaline and noradrenaline to mediate the 'fright-fight-flight' response) and to kidney hormones, which also regulate blood pressure.

So, while high blood pressure in itself is one of the largest risk factors for cardiovascular disease and stroke, there are additional effects of having sleep apnoea. The thin internal lining of blood vessels, the endothelium, has an important function: it detects changes in blood flow and releases substances that regulate the calibre of blood vessels in response. This normal functioning of the endothelium is impaired when blood pressure is high, and this dysfunction is considered an early stage of cardiovascular disease. Experimental models of intermittent hypoxia – recurrent drops in oxygen – suggest that these fluctuations in oxygen themselves lead to endothelial dysfunction, and therefore increase the risk of arterial problems.

Moreover, it is not just the blood vessels themselves that are at risk. Within all our cells, a number of antioxidant mechanisms are in place to protect us from the toxic breakdown products of oxygen metabolism, termed 'reactive oxygen species'. Intermittent hypoxia results in a decrease in these antioxidant mechanisms, and puts our tissues at risk of these reactive oxygen species when the oxygen levels rise again.

As if that was not enough, it seems that sleep apnoea has other effects that compound the risk. While we know that sleep apnoea is strongly associated with obesity, it may be that sleep apnoea actually contributes to weight gain and related problems. The intermittent hypoxia that is a feature of sleep apnoea has been shown to fundamentally influence the effects of insulin, the hormone that controls the breakdown and storage of glucose that is crucial to the prevention of diabetes. Sleep apnoea lessens the body's response to insulin, causing what is termed insulin resistance, the first step in the development of diabetes, and results in higher blood glucose levels.

Sleep apnoea also influences the levels of two hormones called leptin and ghrelin, important for the regulation of appetite and metabolism. So it may be that sleep apnoea in itself increases calorie intake and how those calories are processed, predisposing to further weight gain. What's more, the intermittent hypoxia of sleep apnoea also has profound effects on fat itself, causing inflammation in fat tissues, which may increase the risks associated with obesity.

Thus it seems that sleep apnoea potentially represents a perfect pathological storm when it comes to damage to our blood vessels. The combination of sleep deprivation and episodes of low oxygen gives rise to diabetes, high blood pressure, and inflammation and damage to our blood vessels. The weight gain that it causes just makes everything worse.

Respiratory physicians have been tackling sleep apnoea and its resultant effects for decades, but we neurologists are arriving late to the party. We really only considered it as a diagnosis to exclude in patients in whom narcolepsy was being investigated. I recall as a junior doctor sitting through a presentation on sleep apnoea and being underwhelmed; it was a straightforward diagnosis, with a straightforward treatment, of little relevance to my patients. At the time, the implications of this condition were less well understood, but we now know that sleep apnoea has far-reaching ramifications in the world of neurology too.

Apart from the risk of stroke attributable to high blood pressure and blood vessel disease, sleep apnoea is also a risk factor for an irregular heart rhythm that can predispose clots to form in the heart, before shooting off to obstruct blood vessels in the brain. Sleep apnoea is also of relevance to headaches such as migraines and morning headaches, can result in increases in pressure inside the skull and can worsen the effects of Parkinson's disease.

Within my own clinical practice, it can trigger sleepwalking, sleep paralysis and nightmares, and can complicate narcolepsy and insomnia. Perhaps most importantly, in my epilepsy clinic

I see many patients in whom sleep apnoea complicates their epilepsy control. Sleep deprivation is a potent trigger of seizures in some individuals, and the disruption of sleep by sleep apnoea, as well as the irritation of the brain by recurrent low oxygen, can make the control of epilepsy much more difficult. Recognition and treatment of severe sleep apnoea can make a big difference to some people with this condition.

From the perspective of the general population, one of the most worrying, but also potentially exciting associations, is that of Alzheimer's disease. While sleep apnoea appears to cause significant deterioration in various aspects of cognition, like attention, vigilance, long-term verbal and visual memory, reasoning and problem-solving, it may also have a direct role in the development of dementia. Sleep apnoea appears to hasten the development of cognitive impairment and Alzheimer's, and treatment of sleep apnoea in Alzheimer's causes an improvement in cognitive function. In the elderly, build-up of the protein beta-amyloid has been associated with sleep apnoea. This protein is deposited in the brain, forming plaques that cause signalling problems between neurones, triggering inflammation within the brain, and are one of the microscopic signatures of Alzheimer's disease. And recently, an explanation for this association between sleep apnoea and dementia may have been forthcoming.

When I was at medical school, we were taught about the lymphatic system, a series of vessels or channels within the body that drain fluid from the tissues back into the circulation. Roughly three litres of fluid pass through this system a day, helping to remove waste substances and toxins from the tissues, but also flowing through lymph glands to facilitate the immune system. Damage to the lymphatic system after surgery for breast cancer, for example, can cause lymphoedema, swelling of the arm due to inadequate drainage of this fluid. An infection in the foot may cause lymph glands in the groin to swell, as the immune system kicks in to fight that infection and the lymph node enlarges in response. At the time of learning about this aspect of our bodies, we were also told that the one organ that did not have a lymphatic system was the brain. But, in the past few years – and it is truly remarkable that it is so recent – this received knowledge has been overturned.

Tiny channels within the brain, a network of spaces like the canals draining the lowlands of Holland, have been discovered, and have been termed the glymphatic system. These act like a waste disposal chute for the brain, and there is clear evidence that one of the substances cleared by this network is beta-amyloid. But what has this to do with sleep?

Well, evidence from sleeping or anaesthetised mice shows that the channels of the glymphatic system expand during sleep, and the flow of fluid through this network increases in both these states compared to mice that are awake. In humans, levels of beta-amyloid in the cerebrospinal fluid, where the fluid inside the glymphatic vessels end up, are highest in the morning, suggesting a similar flushing-out of beta-amyloid overnight. In fact, a recent study in humans has shown that even after a single night of sleep deprivation, levels of beta-amyloid in certain parts of the brain, including the hippocampus, often damaged in Alzheimer's, go up. So, if you consider that sleep apnoea disrupts sleep sometimes hundreds of times per night, it is easy to understand why this might impair the functioning of the glymphatic system, why beta-amyloid levels in the brain might go up, and why this might predispose to Alzheimer's.

However, it may not simply be an issue of glymphatics alone. As with damage to the blood vessels, it may be that the recurrent episodes of low oxygen specific to sleep apnoea also have effects on the brain. As with other parts of the body, sleep apnoea may also give rise to inflammation and oxidative stress in the brain, altering processes that maintain the health of neurones and the way in which circuits are regulated.

So, why do I refer to this association as exciting? Well, given how common both Alzheimer's disease and sleep apnoea are, and increasingly so, some researchers have proposed that sleep apnoea is a major modifiable risk factor for Alzheimer's. We cannot modify our genes (yet!), but if sleep apnoea is identified and treated at an early stage, could this be a preventative treatment for Alzheimer's disease?

* * *

Having raised the diagnosis of sleep apnoea with her GP, Maria was referred to our centre. A simple piece of kit that she wore at home at night to monitor her breathing, heart rate and oxygen levels revealed that her oxygen levels were dipping an alarming eighty-six times every hour over the course of the whole night. I ask her how she felt when she received the diagnosis. 'The first reaction, I'm ashamed to say, was that when I was told I had to stop driving, I burst into tears, because I rely on my car for work and my husband doesn't drive. So I was kind of fixated on that. And then, once I'd had time to absorb it, it was a huge relief that it wasn't anything more serious.'

Maria works in healthcare, and I ask her if she had previously come across sleep apnoea. 'I was familiar with it,' she says, but continues:

> I don't think people would necessarily jump to it for a diagnosis for somebody like me. Because I'm a woman, I'm not elderly, I haven't got a particularly large neck.

> I think that's why my GP did not think of it. I don't fit the classic mould of somebody with sleep apnoea. And although doctors asked if I was sleeping okay, they didn't delve into me getting up frequently in the night, things like that.

There are various treatments for sleep apnoea. Obviously, weight loss helps, and for selected patients options include keeping them off their back, oral devices to hold the lower jaw forward and open up the airway, or surgery. More recently, for very serious cases, an electronic device can be implanted in the neck to stimulate the nerve that retracts the tongue in sleep. But at this severity of sleep apnoea, there was really only one immediate treatment choice for Maria – CPAP, or continuous positive airway pressure.

This device aims to prevent the airway collapsing by holding the airway open in sleep. It does this through a mask strapped to the face to create a seal, attached to a small machine that pumps out air under pressure. This pressurised air splints open the airway. Before this device was invented, patients with extremely severe sleep apnoea would occasionally have tracheostomies, surgical procedures to create a hole in the neck to bypass the obstruction. In comparison, CPAP is very minor. But the prospect of sleeping with a mask strapped to your face is not without difficulties. Some people find it claustrophobic or uncomfortable. Maria tells me:

> The first morning I woke up, I felt like I'd been punched in the face, from the pressure in the mask and everything. So I was thinking, *Oh, this isn't great*. But the next day I woke up and it was the first time in eight years that I'd only woken up once in the night. I felt so much better, and

> I just burst into tears because I was just so pleased that this was something that was going to work.

Maria has persisted with the mask and machine, and I ask her what the results of ongoing treatment have been. She tells me her sleepiness has completely resolved, and she is driving again. 'I can focus a lot better. I've been able to start reading, which has been fantastic. I used to just come home, switch on the TV – I'd be asleep within half an hour, so I didn't get to watch anything good. But now I'm able to read and focus on reading an actual novel.'

Also, whereas previously she found herself unable to cope and less able to tolerate minor upsets, she now laughs. 'Some people are irritating and you can't help but react to that. But I feel much more resilient and much more able to cope with things. That part of my personality has returned. I feel much more laid-back and with a bit of perspective on the world.'

I ask her if she feels that the CPAP has unravelled all the consequences of sleep apnoea.

> I don't feel back to my old self. But it is eight years later; I've got two young children, I'm fat and in my mid-forties. So I don't feel the same way as I did [before the sleep apnoea], but I feel much better than I did. I feel optimistic about feeling better in the future. The thought of exercising and hopefully losing enough weight that I won't need to use the CPAP in a couple of years' time is something that is achievable now. A few months ago, it just wasn't at all.

As we chat, Maria expresses something that surprises me. 'I feel horribly guilty about it now, that I didn't listen to [my husband] and take it seriously, and investigate it sooner than I did,' she shrugs. The relief of the diagnosis of sleep apnoea, and its successful treatment with CPAP, has been tainted by regret. She tells me that after she was diagnosed, she researched her condition. 'I read about it. Obviously I had a look on Google like everybody does. I was shocked at how serious it actually was and the damage that I've probably done, having it and not having it treated.' This is not a sentiment I have often heard, but perhaps stems from Maria's background as a healthcare professional.

Maria's regret should be predicated on the basis that treatment of sleep apnoea with CPAP actually does reverse all the negative effects on health. If sleep apnoea really gives rise to all these other health issues, then surely treating it should reduce your risk, of heart disease, stroke, Alzheimer's disease, and so on? What seems like a simple question gives rise to a very complicated answer, however. Benefits of CPAP on cognitive functioning, mood and sleepiness in patients with sleep apnoea have clearly been demonstrated. We also know that CPAP results in an improvement in blood pressure and the endothelial dysfunction that is a marker for blood vessel health, and the effects appear to be more marked in people on multiple drugs for hypertension. But does this translate to a reduction in risk from cardiovascular disease and stroke?

Despite numerous studies, results have been mixed. Some observational studies have suggested that CPAP reduces the risk of death from cardiovascular disease. More recently, controlled trials, where patients have been randomly assigned CPAP or no CPAP, have not borne these findings out. In the field of dementia, CPAP has been shown to result in changes in brain volume, particularly in the hippocampus, a brain area implicated in memory and Alzheimer's, after only a few weeks of treatment, but large-scale randomised controlled trials in dementia have not yet been published.

So why the confusing results in cardiovascular disease? These trials highlight some of the methodological issues surrounding these kinds of studies. Firstly, we know that CPAP compliance – how well people adhere to using CPAP every night for the whole night – is very variable and often limited. In some of these studies, average compliance was very limited, only three hours a night. We don't know how much usage is optimal, but presumably the more you use CPAP, the more benefit might be expected, so it is difficult to draw conclusions on studies where CPAP usage is short. The second issue is the selection of patients for inclusion into these studies. When someone is very sleepy, it is problematic from an ethical perspective to randomise a patient into no treatment. Some of these trials have therefore been performed on non-sleepy patients, and it has been argued that these are the least likely patients to obtain benefit. Thirdly, it may be that some of the damage done by sleep apnoea is irreversible, so the effect of CPAP on modifying future risk is limited. Finally, there remains uncertainty as to how long these trials need to be. Are the reversible consequences of sleep apnoea apparent immediately, after three months, or three years?

It is important to recognise that sleep apnoea is not one entity. There is a range of severity, and some people may be sleepy while others are not. Some people have these respiratory events largely in REM sleep, or solely when they are flat on their back. Some may have suffered from it almost their whole lives, while others may have gained weight and started experiencing it very recently. So the impact of sleep apnoea on health, and the consequences of treatment with CPAP, is not likely to be uniform. It is this variability that probably underlies our uncertainty about its effects and the benefits of treatment.

It is clear, however, as I sit opposite Maria, that the person before me now is not the same person as a few months ago. The grumpy, short-tempered, exhausted Maria that she describes has been replaced by someone smiley, energetic and able to cope with a busy home and working life. 'I don't love the mask,' she says. 'I'd rather I didn't have to sleep with it. But the effect it has had on my life has been so positive. I'm very grateful for it.' For her, the benefits of CPAP, in terms of her sleep, her mood and her function, are evident.

As for the long-term effects, for Maria and others with sleep apnoea, the hope is that future studies will provide the answer.

5

THE SLEEP-TALKING BUS DRIVER

Things are not always what they appear to be. Very early on in a medical education, one learns to be a bit more circumspect about what is presented. When we learn about general practice in medical school, we are taught to seek out 'the hidden agenda', the real reason someone is coming to see their doctor. Is the reason for someone's visit really about their abdominal pain, or is it because they are depressed and have issues at home? Are their headaches the major problem, or are they being bullied at work? As we progress through our medical careers, we become familiar with circumstances where appearances are deceptive: the child in A&E whose bruises are not consistent with falling off a swing, but are more in keeping with abuse; or the elderly man with dementia, admitted with an acute deterioration, who turns out to have been stable for the past few months, but whose family are simply unable to cope with him at home any more. And, as I have seen on a number of occasions, the person with severe intractable epilepsy, uncontrolled despite many medications, who is repeatedly admitted to intensive care so that prolonged seizures can be treated with general anaesthesia, only to find that when scalp electrodes are attached, the brainwaves are totally normal. These intractable seizures turn out to have a psychological rather than a neurological basis.

For a few patients, there is a conscious effort to simulate medical conditions, symptoms and signs for some sort of gain.

A soldier faking an injury to escape arduous duties, or a prisoner trying to relocate from the main prison onto the hospital wing, known as malingering. Then there is the person with Munchausen's, a mental disorder where people invent medical symptoms and ailments in an effort to obtain medical attention, desperately trying to convince doctors that one's abdominal pain is an appendix about to burst, for example. For the vast majority of people, however, this is not 'swinging the lead' or 'putting it on'. These symptoms are very real to them, entirely outside of conscious control; not a malevolent act to pull the wool over the eyes of others.

For centuries, debates about the nature of these 'non-organic' conditions – those that have no basis in biological dysfunction – have raged. Cases of what, in hindsight, represent non-organic neurological disorders have been described as far back as 4,000 years ago in ancient Egyptian texts. In the era of Hippocrates, these conditions were termed hysteria, derived from the Greek for 'womb'. Hysteria was thought by the male-dominated medical world as a condition of women, caused by shifting of the womb out of its normal position, moving around the body to produce a variety of symptoms. It was only at the tail end of the seventeenth century that hysteria began to be viewed as less of a physical disorder, and more of an emotional problem.

Proper study of these disorders really began in the nineteenth century with French doctors like Briquet, Janet and Charcot. At that time, hysteria was still seen as a 'neurodegenerative' disorder, cured by hypnosis. Jean-Martin Charcot would give public demonstrations of patients with hysteria, with seizures, strange movements, paralysis or numbness, apparently successfully treated in front of the eyes of the paying public. There is a famous painting by Pierre Brouillet that features Charcot displaying a hysterical woman to post-graduate students at the Pitié-Salpêtrière Hospital in Paris, then and now a major neurological centre. In the audience, watching in rapt attention, are famous neurologists like Joseph Babinski, Georges Gilles de la Tourette, Henri Parinaud and Pierre Marie, all of whose names are now associated with a variety of clinical signs or disorders in common usage. Reproductions of this painting, *A Clinical Lesson at the Salpêtrière*, have adorned the walls of many of the neuroscience units I have worked in.

However, it is perhaps the work of Sigmund Freud that has had the most impact on our view of these disorders. Working under Charcot, and subsequently in Austria, he viewed hysteria as having a psychological origin, where awful experiences were suppressed and were converted into physical symptoms. Freud essentially described a Catch-22: either you recalled having major trauma, which explained your hysteria, or you repressed it into your unconscious, which also explained your hysteria. This idea of conversion of psychological distress into physical problems led to hysteria being known as 'conversion disorder', and to some extent Freud's views have remained to the present day. But, despite the best efforts of psychiatrists to uncover evidence of psychological trauma, there are patients with this type of problem in whom no such life events can be identified, and it seems that there are in fact multiple reasons people might develop a non-organic disorder.

The latest iteration of these disorders is now 'functional neurological disorder', reflecting our lack of understanding of the origin of these symptoms in many patients, while expressing the view that while the neurological system is structurally normal, there is a disorder of neurological function. A commonly used allegory is that, while the hardware is intact, there is a problem with the software.

Historically, the mind and the body have been viewed as separate entities, termed mind–body dualism, with the origins of human experience in the soul, entirely distinct from 'organic' or physical dysfunction of the body. This concept of dualism is slowly being chipped away, however, and, if you think about it, this should be of no surprise. We know that one's psychological state and neurological function are intrinsically linked. People with epilepsy or migraines will often say that stress is a potent trigger for their attacks. And, on a more day-to-day level, we will all be familiar with the goosebumps we get when scared or excited, the hairs on the back of our neck standing up, our pupils dilating. These physical changes are mediated by our nervous systems, but are clearly under the influence of our psychological state.

On a personal level, a familiar feeling when walking onto the stage to give a big lecture, riven with anxiety, is the sudden sensation that my walking is no longer automatic, but requires concentration. In my nervousness not to trip as I walk up the stairs to the podium, what should be the unconscious act of placing one foot in front of the other becomes a task that needs conscious effort, as my legs become heavy and clumsy. So it seems that our nervous system is vulnerable to our attention being focused upon it; that this awareness of a body part or movement can in itself cause symptoms. In fact, try it for yourself. Sit quietly in a chair, without distraction. Now focus hard on your feet for a few minutes: think about how they feel, the sensation of your socks on your toes, your shoes pressing on your heels or ankles. After a few minutes, you are likely to feel things you didn't feel before, perhaps a tingling, a numbness or a little discomfort. The mere act of attention can generate symptoms. It is likely that for some patients with non-organic symptoms, especially those without an obvious psychological trigger, it is this mechanism that causes their problems.

Moreover, these non-organic neurological symptoms are amazingly common. Some studies have suggested that functional problems are the second most common reason for patients coming to a neurology clinic, after headache.

Of course, it will come as no surprise that these cases also occur in the world of sleep medicine. I regularly see patients who are excessively sleepy as a manifestation of severe depression, or others who present with sleep attacks that are an unconscious mechanism of escaping from stressful situations. I can think of several patients who have been referred to me with apparent neurological disorders of sleep that originate from some kind of childhood emotional, physical or sexual abuse occurring at night, in the bedroom. I have seen more than one patient with apparent collapses, previously diagnosed as part of narcolepsy, in which the ultimate diagnosis was one of a functional neurological disorder. As with almost all the patients I see in my general neurology clinic, these patients had no conscious awareness of the underlying nature of their symptoms.

But I also remember one of the first cases I saw as a new consultant, of a young man who had developed a profound and untreatable sleepiness shortly after the birth of his child. For the past five years, he had missed out on his son's childhood, largely confined to bed, sleeping twenty hours a day. His partner was essentially bringing up their son as a single parent, dealing with the daily rigours of childcare while also trying to hold down a job. For the previous few years, neurologists elsewhere had tried a variety of treatments to keep him awake, to no avail.

However, a prolonged admission revealed that for a large

chunk of the day he would lie in bed 'asleep', while the electroencephalogram – analysis of his brainwaves – clearly showed him to be awake. After a struggle to convince him and his partner that the issue might be psychological rather than neurological, I arranged for him to be admitted for four weeks to a neuropsychiatric unit for assessment and treatment. Three days before he was due to be admitted, I received an email from his partner. 'It is amazing,' she told me. 'He has suddenly woken up and is so much better! For the last few days he is wide awake and engaging with family life, and is back to his old self.' I can only conclude that having avoided his family responsibilities for several years by lying in bed, he was concerned that he would not be able to maintain the façade under the close observation of an in-patient unit.

Doctors, or almost all doctors, are driven by the desire to help patients, to be on their side. It is against our nature not to believe what we are told, and sometimes it is difficult to maintain the degree of cynicism necessary to get to the right diagnosis. Too much scepticism is undoubtedly worse than too little, but getting the balance right can sometimes be extremely challenging. I have definitely been caught out a few times in my practice.

* * *

Robert is also not what he appears to be. Now seventy-two years old, I first met him six years ago. He is tall, slim, with close-cropped grey hair; confident and erudite. He talks with a crisp, cut-glass, upper-class accent. He tells me of his youth, how he was born in a leafy part of central London, of glorious summers with his extended family in Ireland, and of his schooling at a good academic institution. Despite this, he left school at sixteen, eager to work but without clear guidance or focus, and coasted between the music industry, marketing, publishing and various other jobs, quickly getting bored and keen to dip his toe into other walks of life.

'I had a very low boredom threshold, hadn't a clue what I wanted to do, had absolutely zero direction,' he says. At one point, he tells me of his youthful plans to get into politics. I ask him which party he wanted to represent. He laughs as he tells me: 'The women were much more attractive in the Conservative Party, so I joined them!'; later in our conversation, he admits that nowadays he would be described as 'very politically incorrect'. In fact, he went on to set up a branch of the Young Conservatives in a very affluent part of London, before once again getting bored and moving on.

He strikes me as very much a member of the Establishment, and, given his background, I am puzzled as to how he ended up working as a bus driver. 'Now there is an interesting story. If someone had told me I should drive a bus when I was in my twenties,' he says, 'I would have punched them.'

After floating around London for a few years, he went off to the United States. 'I went there in the early to mid-'60s. It was quite eye-opening,' he tells me, with a twinkle in his eye. I ask him what he did there. 'As many drugs as I could possibly lay my hands on!' he replies, deadpan.

Eventually he drifted back to the UK, where his interest in illicit drugs served him well, and he became a drug dealer. I tell him that his story is reminiscent of Howard Marks, 'Mr Nice' – a notorious drug smuggler whose cannabis empire germinated from selling the drug to friends and acquaintances during his Oxford undergraduate days in the '60s. 'Oh, yes, I knew him well!' Robert chimes. 'I believe I am in the book,' he says, referring to Marks's autobiography, published a few years ago.

Like Marks, Robert's career culminated in him serving time

at Her Majesty's pleasure – 'Well, the government decided that they thought that something that grows quite naturally is illegal. And because I brought a large quantity of it into the country, they decided the best thing to do was to stop me from bringing any more of it in. I got done for the importation of cocaine,' he tells me.

Robert was sentenced to nine years, serving out the last year of his sentence in an open prison. At that time, in his fifties, he was allowed to work, but unsurprisingly his options were limited, hence his current occupation of driving buses. As his story unravels, I am forced to rapidly re-evaluate my view of him.

When he tells me why he is at the sleep clinic to see me, though, I am torn as to whether it is his life story or his sleep problem that is more extraordinary. For the past few months, Robert has been talking in his sleep. It is not so much the fact that he is talking that is the issue, he says, but the precise nature of *what* he says.

It all started a few weeks after moving in with his partner of eighteen months, Linda*. 'She says I talk constantly in my sleep,' he tells me. There is no previous history of sleep-talking or sleepwalking, so Robert is somewhat surprised by this revelation. Linda's objection, though, is focused primarily on *what* he is saying. 'Apparently I was talking about my love for a previous girlfriend, Joanna. It's very odd, as I haven't been in contact with her for years.'

Robert is mystified, as of all his previous girlfriends, Joanna would not be the one he would declare undying love to. On one occasion, Linda says that he recited Joanna's phone number in his sleep. For the life of him, Robert would not be able to remember her phone number in waking life. 'I thought she was completely nuts because, if I had been waterboarded, I couldn't have remembered that number,' he recalls. To his astonishment, Linda checked the number in his old address book and found it to be absolutely correct. Needless to say, this was causing some friction in the relationship, but not as much as subsequent events.

As time has progressed, Robert will be woken on a regular basis by Linda screaming at him, telling him of the unspeakable things he has been saying. 'Apparently, I have begun talking about abusing Joanna's family, her kids, even her parents – males and females. I have been talking about necrophilia, even bestiality!'

His sleep-talking has obviously led to frequent arguments, and he has been spending an increasing number of nights in local hotels or sleeping at his sister's or friends' homes. Shocked by this turn of events, Robert has already been to see a psychiatrist, worried that he might have something seriously wrong with him. Given the all-clear, he is concerned that he may have a neurological problem – a brain tumour or dementia – that may be causing this dramatic change in his night-time activities.

On hearing the story, I am taken aback. Having seen many patients with unusual behaviour at night, I have never come across anything like this. It is unusual on many levels. Sleep-talking is almost always a variation of sleepwalking, a non-REM parasomnia, common in children but rarer in adults. Like Jackie's sleep-motorbiking and Alex's night terrors, this type of sleep-talking arises out of the deepest stages of sleep, when parts of the brain are awake while the rest of the brain remains in very deep sleep. People can have full-blown conversations, with their eyes open, and look to all intents and purposes awake, but the chatter will often be inappropriate to the circumstances and will relate to some situation the sleep-talker imagines themselves in. Robert's story does not sound typical of this, however.

Firstly, he is sixty-seven years old when we first meet, and to suddenly start having a non-REM parasomnia at this age, with no suggestion of anything earlier in life, rings alarm bells. But it is also the nature of the conversation that is unusual, to say the least. Most people will talk of mundane matters, occasionally an argument, or, in the case of sleep terrors, some impending disaster. To have such a persistent and narrow subject of conversation, with these very dark undertones, is entirely new to me.

The other possibility is that he is suffering from REM sleep behaviour disorder, the condition that John in Chapter 3 suffers from, where the paralysis that accompanies dreaming sleep fails to materialise. This often comes on later in life. But in REM sleep behaviour disorder, the speech is rarely intelligible, and is more usually swearing, shouting or mumbling. And if this were to be REM sleep behaviour disorder, I would expect Robert to have some recall of his dreams when woken by Linda shouting at him.

Somewhat at a loss, I suggest to Robert that we bring him in for a sleep study, to observe what is going on overnight, and that I meet Linda to get more details. Robert is happy to be admitted for the night but is more negative about my meeting Linda. She is not keen to discuss their issues.

A few weeks later, I meet Robert again to discuss the results of his sleep study. It does indeed show something. He has moderately severe sleep apnoea. He snores loudly and stops breathing about thirty times an hour. But there was no sleep-talking. Apart from the rumblings of his snoring, there was nothing. Not even a whisper or a groan. It is common not to see evidence of a non-REM parasomnia on a sleep study, but Linda reports him doing it every night. At least I am able to exclude REM sleep behaviour disorder; the expected paralysis in REM sleep is visible on the study. Nevertheless, I am only a little further forward in trying to establish a diagnosis.

One of my senior colleagues, tongue firmly in cheek, frequently uses the adage, 'If in doubt, pressurise the snout.' This harks back to the days when sleep medicine was the sole preserve of respiratory physicians, who saw everything through the prism of sleep apnoea, the collapse of the airway during sleep. It refers to the use of CPAP, the mask delivering pressurised air to the airway that Maria found so helpful. But in this context, this course of action makes sense. I know Robert has significant sleep apnoea, and in theory the disruption of his sleep by thirty breath-holding events every hour could be triggering non-REM parasomnias. So, in desperation, I organise a mask and machine for him to take home, fingers crossed that he comes back saying that he is cured of his sleep-talking, hoping that he comes back thankful that I have heroically saved his relationship.

A couple of months later, I see Robert again. I ask him how everything is going with the CPAP. He says that his sleep is a little better. He is feeling a bit more refreshed on waking. But what about the sleep-talking, I ask him? 'Oh, that has been sorted out!' he replies, and I get ready to pat myself on the back for successfully treating him. Until he goes on, that is.

'After the sleep study, I got very suspicious,' he continues, 'and I went down to Argos to buy a voice-activated recorder.' He then proceeds to tell me that, one afternoon, he had a nap on the sofa, wearing the CPAP mask tightly clamped on his face. He was rudely awoken by Linda screaming at him, telling him that once again he had been talking about sexually abusing members of Joanna's family. Having had the recorder on, when he listens back to the tape, he can hear the gentle hum of the CPAP machine, suddenly interrupted by Linda's shrill voice.

'I was asleep in a different room, I had the recorder on, then I was woken by her screaming at me, "Stop talking about Joanna, blah, blah, blah, blah, blah, blah," a five- to ten-minute complete rave, and I had the machine beside me.' On the recording, however, there was no mention of Joanna, abuse, necrophilia or bestiality. In fact, there was no talk at all. My jaw drops to the floor.

* * *

By its very nature, people with sleep disorders are limited as to what they can tell you. Cut off from the external world, in a state of unconsciousness, there is no or limited recall. People can speak of their experiences when they wake, and how they feel during the day, but if you are asleep, you are offline. Similarly to situations like major epileptic seizures, sleep is a brain state defined by reduced awareness or lack of consciousness, and so I rely heavily on reports of bed partners to achieve a diagnosis. And I always assume that the bed partner's story is reliable. Wrongly, as it turns out – at least on this occasion.

So how do we explain Linda's fictitious reports of Robert's sleep-talking? It is almost incomprehensible. But there are a few possible explanations. The first is that Linda is frankly psychotic, that she is hallucinating or delusional, and that these symptoms, her hearing Robert saying these things, is very real to her. Perhaps this represents morbid jealousy, a psychological disorder where someone is profoundly obsessed that their partner is being unfaithful. This condition, also known as Othello syndrome, often presents with constant accusations of infidelity, examining belongings such as diaries, clothes and emails, preventing one's partner from social contact, and occasionally escalating to violence. This can be a form of obsession, but can also be a delusion, a belief without resistance, held to be true despite all evidence to the contrary. It is often associated with other underlying psychiatric illnesses or personality disorders.

Another possibility is that of Munchausen's by proxy. As mentioned before, Munchausen's syndrome represents the disturbing behavioural pattern of sufferers repeatedly presenting to healthcare professionals with elaborate and dramatic health complaints, often going to multiple hospitals under several pseudonyms. Patients will often undergo multiple operations or procedures, will be prescribed many medications, but when efforts are made to obtain background information, they will often become very obstructive, or will then move on to the next hospital or doctor. Sometimes patients will fake symptoms using drugs. I have seen patients putting eye drops in to dilate their pupils, faking neurological problems; and there are case reports of patients injecting egg protein into their bladder to simulate kidney problems, or injecting themselves with insulin to cause a coma. This is thought to be a form of radical attention-seeking behaviour and originates from a personality disorder. In itself, Munchausen's syndrome, named after Baron Munchausen, a literary character known for his dramatic and elaborately tall tales, is incredibly rare. But Munchausen's by proxy is even rarer.

Rather than simulating symptoms in themselves, perpetrators with this condition malevolently create symptoms in others, once again in an effort to gain attention. This is a sinister and dangerous form of abuse. Mortality rates are as high as one in ten, due to poisoning or complications of unnecessary medical interventions. By its nature, the victims of Munchausen's by proxy are vulnerable and dependent on the perpetrator, unable to report their own symptoms, incapable of preventing poisoning or sabotage of investigations, and so are almost invariably children. But, as with a child, a sleeping adult is largely dependent on someone else to report their problems, making sleep disorders one of the few scenarios in which an adult may be a victim of this horrifying abuse. However, Linda has never appeared in front of me, and seems reluctant to interact with the medical profession in general, making this diagnosis rather unlikely.

A few months later, Robert tells me he has managed to convince Linda to see a psychiatrist herself. Unbelievably, she has been given a clean bill of mental health. In my own mind, I wonder if what this represents is gaslighting.

This term describes another rare form of psychological abuse, where perpetrators seek to manipulate their victim by making them doubt their own sanity, memory or perception. The end goal may simply be about control, or destroying someone's self-esteem, but can be for financial gain or to engineer the break-up of a relationship. Its name derives from a 1938 play, subsequently made into two films in the 1940s, one with Ingrid Bergman, all called *Gaslight*. A husband tries to convince his wife that she is slowly going insane, aiming to have her institutionalised and get her out of the way, so that he can search for jewels hidden in their house. He isolates her, fakes evidence of her stealing small objects, and engineers jealousy by flirting with the maid before telling his wife she is imagining it. During his search for the missing jewels, his actions cause the gaslights in the house to flicker, but he convinces his wife that these changes in the gaslights are figments of her imagination. I ponder if Linda may have been trying to engineer a split or manipulate Robert for some reason, and, indeed, shortly after the case comes to a diagnosis, we write up Robert's case in the medical literature as possible gaslighting.

* * *

As with so many aspects of this case, I am later forced to reassess yet again. I lose touch with Robert for quite some time, only to see him again five years later. With the passage of time, it has become clear that Linda's behaviour suggests more than gaslighting. Rather unsurprisingly, the relationship came to an abrupt end shortly after I last saw him, but Robert remains tormented by her. He has had to go to court several times to seek banning orders against her. She has called his workplace and the police several times, making multiple allegations against him, sexual and criminal. She has been highly manipulative of the police and of the courts. On one occasion, he was pulled over by the police while driving his bus. Apparently, she had called them and accused him of dealing drugs from behind the wheel of his bus. He was soon allowed to go on his way. Clearly, these are not the actions of someone in their right mind, but nor are they the actions of a gaslighter. I question whether Linda's original psychiatric evaluation was correct.

In retrospect, Robert now admits that he was already a bit suspicious before we first met.

> If she had just stuck to me talking about ex-girlfriends, it might have been believable. If she had talked about me having an affair with the neighbour next door, then maybe. But when she started going on about the necrophilia, the bestiality, abusing whole families of either sex, I thought she was away with the fairies. I might be a lot of things, but homosexuality has never had any appeal to me.

However, he was credulous enough to try some of his own remedies. 'Obviously I tried getting really, really stoned. With a fair amount of booze to go with it. I tried various other drugs that I could get hold of easily. Whatever the latest sleeper was.'

And there clearly remained enough doubt for him to seek more definitive proof. The normal sleep study supplied him with it. He tells me: 'The sleep test was the first time I had anything I could hang my hat on.' I ask him if there were any signs of mental imbalance earlier in their relationship. 'She was an ex-girlfriend of one of my friends. Charming, pleasant, superb cook, and apparently sane,' he tells me. There were no hints as to what was to come. Robert now thinks that she is 'just mad'. While he says she has been very scheming subsequently, he now thinks that she genuinely believed that he was saying what she was reporting, that it was not an effort on her part to control or manipulate him.

I imagine myself in his position. In its most generous interpretation, he has suffered due to the mental illness of his partner, but potentially he has been victim to a gut-wrenching evil ruse. I say that it must have been a very traumatic experience for him, but he is rather phlegmatic. 'More of an annoyance, really. It was a bore. In my life, I have had to deal with a lot, and I simply get on with it. She is just a mad woman.'

* * *

Lessons in the world of medicine come in unexpected forms. Getting a correct diagnosis is always rewarding (and for neurologists, the rarer the better), but it is your mistakes as a doctor that are the most valuable. Robert's case has taught me caution, to not always take information at face value, and is a reminder of the complexities of human relationships. And while I always keep in the back of my mind that what I am seeing in my clinic is a disorder that may have a basis in my patient's psychological state, it does not often occur to me that it might actually be the partner's psychological health that I need to be considering. I suspect that this did not feature strongly in Freud's or Charcot's diagnostic thinking either.

It also illustrates our vulnerability when we are asleep – not only as a result of the physical peril of being oblivious to the world around us, but our susceptibility to our nearest and dearest. Sharing a bed with someone is an act of deep trust.

6

WEAK WITH LAUGHTER

'This 39-year-old patient came to see me complaining of a strange symptom,' reads the referral letter. 'For the last two years or so now, every time he laughs, and sometimes after violent exercise, he at first feels a strange sensation in his back, followed by a feeling of weakness, and then he suddenly loses his balance and falls down.' As I read on, the possible causes for the collapses that the patient, named Adrian, experiences go through my mind. Could this represent epilepsy, a faint or a sudden change in cardiac rhythm? Could this represent some damage to the spinal cord? The detail that catches my eye, however, is the reference to laughter.

When Adrian walks into my general neurology clinic, I see a man roughly my age. He is a little shorter than me, with greying hair, smartly dressed in a suit, having come straight from work. He is friendly, open and humorous, but clearly puzzled as to what on earth is going on. He sits down in the chair opposite me, and begins to tell me his story.

He recalls the first time he suffered from one of his collapses. Adrian had driven to see his parents. 'I had crept up the driveway, where I could see that my mother was in the back garden doing some gardening,' he recollects. 'She was next to the garden fence, so I thought it would be really quite amusing to sneak up and hammer on the fence. She would jump and everyone would laugh. Which is what duly happened.' But that was not all. 'What I didn't expect at the time was that I completely slumped to the floor, slowly down the garage doors, and ended up lying in the driveway for I guess fifteen to twenty seconds with a slight case of the shakes.'

He goes on to describe further events.

'There were a couple of [other] instances – one at a friend's fortieth birthday party. Admittedly, I'd had a couple of beers, but it wasn't the end of the evening, and I was perfectly capable of standing. So I was standing next to a low sofa and friends were sat on that sofa. And there's me being amusing, and I promptly ended up lying across my friends on the sofa. I just ended up in their laps.'

On one occasion he was on holiday with his family, and they were sitting at a picnic bench while his daughter was eating some biscuits. 'So we were sat there – again me being amusing, so I thought,' he chuckles. 'I said whatever it was I said and ended up wiping her Jaffa Cakes off my face, because I just slumped forward and ended up with a face full of cake.'

We talk in more general terms about his collapses. It is clear that he does not lose consciousness in any way. He remains fully aware throughout, and can hear everything and see everything. It is almost as if someone has simply pulled the electric plug. All of a sudden, he loses strength in his body, his muscles go weak, and he slumps to the ground like a rag doll.

> It starts off with a sort of stirring feeling. This all happens very, very quickly, and I've learned to catch it sometimes. Then [there is] almost a weakness through my legs and my lower back in particular. The strength in the muscles just seems to evaporate. And it doesn't happen instantly, as if you were fainting or something like that. It's a slow collapse. It's not something typically

> that can be stopped. I just sort of slump to the floor. And I use the word 'slump' rather than 'fall' and 'collapse' because it's a slow process.

I ask him what happens at the end of an attack. 'It eases a bit. But it's not like all of a sudden you spring back into life. It's more of a coming around.'

The trigger for these events is clear, as his doctor reports in the referral letter. Adrian comments: 'The trigger is usually when I am – it sounds big-headed – but [when] I am the source of the humour. I'm telling a joke or I'm doing something [funny]. I can sit and watch a comedian, I can laugh myself inside out without any effect at all. But if I were to say something that I felt was very funny to you, there's a good chance that I would end up on the floor. So it's definitely self-inflicted, it seems.'

When I later meet Adrian's wife and daughters, they tell a similar story. His two daughters, Sophie, aged thirteen, and Erin, aged ten, discuss his sense of humour. In response to me asking if their father is funny, Erin responds with a definitive 'No!', as I am sure my daughters would say about me. Sophie is marginally more generous. 'Sometimes,' she says. 'It really depends on the context of the joke. Obviously me being older, I have a more mature sense of humour than Erin, so sometimes they are funny. But otherwise, no, not at all.'

Sophie then proceeds to giggle her way through various examples of her poor father's collapses. She and her sister describe a family trip to the zoo. 'We had gone to the monkey sanctuary. We were walking down past an enclosure. Obviously thinking he was hilarious, Dad saw a sign that said "blue tit" on it, as in the bird. He made an inappropriate joke.' (At this point I can see Adrian's wife behind the two girls desperately trying to suppress a fit of laughter as the two girls cackle at the memory.) Sophie continues: 'He slumped over the fence [in front of the enclosure], which was quite alarming. I don't know what the people walking past us thought was going on. Half his body on one side, half on the other!'

* * *

As a medical student, I relied in part on a textbook called *Neurology and Neurosurgery Illustrated* to revise for my final exams. As I leaf through it today, I cringe, and am struck that this is essentially the equivalent of *Neurology for Dummies*. As I look more closely, however, I see that the spine of the book is broken at page 103, and I remember being fascinated by its contents.

This single page details a condition known as narcolepsy, a neurological disorder poorly understood at that time. It describes the irresistible desire to sleep in inappropriate circumstances and places, sleep paralysis, hallucinations at the point of drifting off to sleep, as well as cataplexy – 'the sudden loss of postural tone. The patient crumples to the ground. Consciousness is preserved. Emotion – laughter or crying – can precipitate an attack.'

I recall being amazed that a neurological disorder could manifest in such a myriad of bizarre and surprising ways, and being drawn again and again to this page. I am thankful that my knowledge of narcolepsy is no longer limited to the contents of this book, and as I talk to Adrian, it is very obvious that he is describing cataplexy.

He tells me: 'My arms will droop, my head will droop, and I usually get a bit of a quiver going on. But I'm able to hear, I'm able to see, breathe obviously. There's full consciousness, but you're essentially paralysed, you're unable to move. And that will vary in length from five to thirty seconds. It depends. But it's almost always a complete collapse.'

Confusingly, despite him working long hours in the financial sector, commuting up to four hours a day, and being profoundly sleep-deprived as a result, he shows absolutely no other features of narcolepsy – no sleepiness, no hallucinations, no sleep paralysis. Although cataplexy is essentially 'pathognomonic' for – very specific and diagnostic of – narcolepsy, there are no other hints that he has this condition. I tell him that, while it sounds like classic cataplexy, we need to do more tests. I organise a blood test, MRI scans of his brain and spine, and a sleep study to look for the characteristic signature of narcolepsy.

* * *

So what is narcolepsy? And how does it cause these strange phenomena? At the time of my medical school finals, this disorder remained much of a mystery. Over the past twenty years, however, our understanding of this disease has been completely transformed. In my mind, this is about as pure a neurological disorder as can be, related to the damage of only a tiny number of neurones deep in the centre of the brain. But the path to this conclusion, the identification of the key to narcolepsy, is intriguing.

The earliest descriptions of narcolepsy with cataplexy date back to the latter half of the nineteenth century, and the term narcolepsy – from the Greek for 'to be seized by sleep' – was first coined by a French physician called Gelineau in 1880. Though it was the epidemic of encephalitis lethargica after the First World War, a syndrome thought to have been triggered by the influenza epidemic sweeping the globe, that caused renewed interest in narcolepsy.

Encephalitis lethargica is still a condition that is much debated, but some researchers think that in susceptible individuals the influenza virus triggered an autoimmune response – the body's immune system directing an attack against itself – that caused damage to parts of the brain controlling wakefulness and movement, causing profound sleepiness and features similar to Parkinson's disease. (Oliver Sacks's book *Awakenings* details the human cost of encephalitis lethargica and the heartbreakingly transient improvements that treatment with anti-Parkinson's medications had.)

It was a Romanian-Greek psychiatrist and neurologist called Constantin von Economo who identified that, in patients with encephalitis lethargica, the sleepiness seemed to be related to damage in an area called the posterior hypothalamus. He was impressed by the similarities between some aspects of encephalitis lethargica and narcolepsy, in which no such damage could be found. This led him to propose, with remarkable acumen, that the changes in the brain in narcolepsy were in the same area, the hypothalamus, but that these could not be seen with the naked eye. 'It is very probable, though not proved, that the narcolepsy of Gelineau, Westphal and Redlich has its primary cause in an as yet unknown disease of that region,' he wrote in 1930.

As is so often the case, identification of the cause of the narcolepsy came in large part from studying animals. In the early '70s, narcolepsy with cataplexy was identified in dogs. Type 'dog' and 'narcolepsy' into YouTube, and you will be faced with hundreds of videos of Alsatians, Dalmatians, Labradors and terriers all dropping to the floor when greeted by their owners with a bowl of food, their tails trying to wag through the complete loss of muscle strength their cataplexy brings. Researchers in Stanford began a breeding programme for dogs with narcolepsy, and by the mid-'70s they had succeeded in establishing that narcolepsy was a genetic disease in Dobermans and Labradors. But it was not until advances in the field of genetics in the 1980s and '90s that, in dogs at least, a region of the genome was identified as being causative of narcolepsy. Subsequent painstaking analysis of the genes in this region led in 1999 to the identification of mutations in a gene of uncertain function – later named 'hypocretin receptor 2 gene' – that appeared to cause narcolepsy in Labradors and Dobermans alike.

I cannot stress enough how intensive this sort of work is. I am still traumatised by my PhD, which involved similar work, but I had the advantage of robots to pipette thousands of samples into vials, and the benefit of working at the Sanger Institute in Cambridge, one of the world's largest centres involved in sequencing the human genome, where much is automated. At that time, it must have been infinitely more difficult.

It was almost simultaneously, by coincidence, that the substance acting on these receptors, hypocretin (also known as orexin), was identified by a team studying genes expressed in the hypothalamus, the very area of the brain proposed by the prescient von Economo as being involved in narcolepsy. We now understand that the neurones producing hypocretin are located in a tiny area of the lateral hypothalamus, but are projected widely throughout the brain. In particular, they link directly to areas of the brainstem involved in the regulation of sleep and wakefulness as well as dreaming.

Narcoleptic dogs do not give us all the answers, however. Very few of my patients with narcolepsy report a strong family history of the condition. The risk of a first-degree relative – sibling, parent or child – of someone with narcolepsy also developing the condition is up to a hundred times greater than that in the general population, but this still means that only about 2–10 per cent of relatives will also have narcolepsy. This is a long way away from what would be expected in a disease caused by a single gene. It implies that, while there may be a genetic component that influences developing narcolepsy, genes in humans are not the whole picture. Indeed, the only human ever to have been found to have mutations in the same gene as our Dobermans is one individual whose narcolepsy manifested at six months old.

Subsequent laborious efforts have shown that, in humans with narcolepsy, hypocretin is without doubt involved, but it is the nature of its involvement in humans that differs from dogs. In the Stanford dogs, there is a genetic error in how receptors for hypocretin are put together; in humans, it is an absence or deficiency of the chemical itself. While the receptor is normal, there is little or no hypocretin to bind to it. Both types of defect result in underperforming hypocretin-mediated circuitry, and cause narcolepsy. But in humans with narcolepsy, particularly in those with cataplexy, the neurones producing hypocretin are damaged. On average, people with cataplexy have lost 90 per cent of their hypocretin-producing neurones. Something causes them to disappear, to simply vanish. And measuring hypocretin in the cerebrospinal fluid – the liquor that bathes the brain and spinal cord, extracted through a spinal tap or lumbar puncture – shows that patients with narcolepsy and cataplexy almost invariably have absent or very low levels.

* * *

Adrian is admitted into hospital for a sleep study shortly after our first meeting. One of the diagnostic tests for narcolepsy, the sleep study, involves Adrian being attached to a variety of sensors, monitoring his breathing, heart rate, muscle activity and brainwaves. Overnight, he sleeps as normal a night's sleep as is possible in this unusual setting. The following day, he undergoes a test called the mean sleep latency test. At two-hourly intervals, the lights are dimmed, and Adrian tries to sleep. He is monitored to see how quickly he drifts off, and what stage of sleep he goes into during these twenty-minute opportunities.

It was the discovery of REM sleep by Aserinsky and Kleitman in the 1950s, described in Chapter 3, that began to give us some insight as to what might cause many of the symptoms of narcolepsy. As we've heard, normal REM sleep – the stage of sleep when we dream our story-like dreams – is accompanied by paralysis of almost all of our muscles. While our brains are almost as active as in waking, our bodies are disconnected. Typically, REM is a stage of sleep that we do not enter into until we have been asleep for sixty to seventy-five minutes. Dreaming in short naps is rare, and dreaming as we drift off to sleep equally so.

In narcolepsy, however, this is not the case. The neurological mechanisms preventing you from entering REM quickly after going to sleep go awry. People with narcolepsy can flick into REM very early in the night, sometimes directly from waking. Multiple episodes of REM sleep in the mean sleep latency test, with its brief, twenty-minute opportunities to nap, are characteristic of narcolepsy. And so it is easy to see why some of the symptoms of narcolepsy arise. If you go straight into dreaming sleep from being awake, then having dreaming experiences in wakefulness may manifest as seeing or hearing things in the room as you lie there. If the muscle weakness of REM sleep is switched on while you are awake, then you will feel paralysed. And if you are really unlucky, you may experience both together, resulting in the really unpleasant sensation of feeling pinned down to the bed while you hallucinate that someone or something is in the room with you.

Imagine two seesaws in a children's playground. When swinging free, balanced without any children on either end, a strong gust of wind may swing the beam one way or the other. Think of one seesaw controlling the switch between sleep and wake, and think of the other controlling the switch between REM and non-REM sleep. The hypocretin system is the equivalent of a 16-stone prop forward, perched delicately right at the far end of each seesaw, firmly planting the end that pushes wake into the ground, and when you are asleep, holding down the end that promotes non-REM sleep. Lose the prop forward, and the seesaws go back to swinging in the breeze. Without hypocretin, people with narcolepsy flick in and out of sleep uncontrollably, and switch in and out of REM. So the loss of hypocretin-producing neurones results in sleepiness during the day, sometimes irresistible sleep attacks, sleep paralysis and hallucinations on sleep onset, termed hypnagogic hallucinations.

In addition, flicking in and out of REM sleep into wake overnight results in extremely vivid dreams. Most people only remember their dreams when they wake from REM sleep, but for people with narcolepsy, they are doing this constantly throughout the night. And actually, although we consider people with narcolepsy to be very sleepy, over a 24-hour period they don't really sleep any more than anyone else. They sleep more during the day, but overnight their sleep is broken and of poor quality.

When I look at the results of Adrian's study, however, searching for the early onset of REM overnight, or the short time to drifting off to sleep and the presence of REM sleep in the daytime nap test, there is nothing. No hint of narcolepsy. His sleep study, apart from showing a bit of snoring and some leg movements, is totally normal. Cataplexy usually develops alongside sleepiness or up to several years afterwards. It is rare for cataplexy to start before any other symptoms of narcolepsy.

When I see him next, we discuss these results. I tell him that a positive sleep study is often seen in people without narcolepsy, and can be caused by sleep deprivation, but a negative sleep study is rather against a diagnosis of narcolepsy. I have occasionally seen apparent cataplexy in people with psychological issues, a manifestation of psychological distress, but Adrian strikes me as totally stable, without any obvious psychological triggers. We decide to proceed to a spinal tap, to analyse his cerebrospinal fluid for hypocretin.

Two weeks later I see him again in clinic with the results. Normal. I am slightly stunned. Pretty much everyone with cataplexy has low or deficient levels of hypocretin. We discuss his symptoms again. I remain totally convinced that what he is describing is cataplexy, and despite the nagging doubt in my mind raised by the negative results, I prescribe him medication specific for it.

* * *

While it is easy to understand how a deficiency of hypocretin can cause sleepiness, hypnagogic hallucinations and sleep paralysis, explaining cataplexy is a bit more difficult. Why this bizarre symptom occurring in the middle of the day should be related to a sleep disorder is incomprehensible at first glance. But there are a few clues. Monitoring muscles during cataplectic attacks demonstrates repeated brief silence in muscle activity, and looks very similar to the silencing of muscle tone seen in REM sleep. It is this repetitive nature of short bursts of weakness that people sometimes describe as a juddering to the ground. The weakness usually builds up gradually, and people can often manage to lower themselves to the ground to avoid injury. If seated, the muscles of the neck may briefly sag, causing some nodding as the person tries to combat the muscle weakness, and if the face is involved, there will be a slackening of the facial musculature that looks like twitching.

Another feature of cataplexy that gives us some insight into what causes it is that, occasionally, particularly with prolonged attacks, people begin to hallucinate or dream. It therefore appears that cataplexy is closely linked with REM sleep. In the same way that REM sleep, or at least aspects of it, are switched on inappropriately early in the night in people with narcolepsy, it seems that the absence of hypocretin during the day predisposes to the sudden onset of aspects of REM sleep in wakefulness. The paralysis of REM sleep is switched on while you are standing or sitting. In many ways, this is the mirror image of John's acting out of his dreams at night, his REM sleep behaviour disorder. Whereas John's problem is down to a failure to generate paralysis in REM sleep at night, Adrian's is due to the uncontrolled switching on of paralysis in the day. In Adrian's case, the cataplexy affects his whole body, but for some people it can be limited to only one part – the face, neck, arms or legs, for example.

So, cataplexy results from aspects of REM sleep, and the paralysis that accompanies it, being switched on at inappropriate times during wakefulness. But what all of this fails to explain is why cataplexy is triggered by laughter. In fact, it is not just laughter that triggers it. Sometimes it arises spontaneously, without any prompts at all. For other people, surprise, anger, sorrow or anxiety spark it off. I have had patients collapse in the street when honked at by a car; others who have attacks when they argue with their family. But certainly laughter, or an 'internal feeling of mirth', as Adrian puts it, is the most common generator.

Actually, mild muscle weakness with laughter is a normal phenomenon, hence the expression 'weak with laughter'. Monitoring muscle electrical activity in normal individuals demonstrates that laughter suppresses what's called the H-reflex, essentially a laboratory version of the reflexes elicited by neurologists when we tap the knee with a hammer. Clearly, in cataplexy, this suppression is dramatically amplified. In fact, during a cataplectic event, the knee jerk and other reflexes simply disappear. On the rare occasions I see a cataplectic event in a patient in clinic, if a tendon hammer is within reach I will check for these reflexes to confirm that this is true cataplexy.

Experiments have shown that the hypocretin-producing neurones in the hypothalamus are very active when we experience strong emotions. So it seems that somehow hypocretin puts the brake on this normal phenomenon of weakness generated by strong emotions, dampening it down. The absence of these neurones somehow destabilises the brainstem's regulatory system for muscle tone.

Though that is not the whole story either. In addition, another part of the brain called the amygdala is implicated. This almond-shaped structure sits on either side of the brain, deep within the temporal lobe, and has an important function in the processing of emotional stimuli. Epileptic seizures arising in the amygdala often precipitate the experience of sudden strong emotion, such as overwhelming fear. Studies in patients with narcolepsy have shown alterations in activity in the amygdala while looking at funny pictures, and in narcoleptic dogs there appear to be changes in electrical activity in the amygdala with cataplectic attacks. The theory is that circuits from the amygdala project to the areas of the brainstem involved in maintaining muscle activity. In wakefulness, these connections are inhibited by hypocretin, but in narcolepsy, the lack of hypocretin takes the brake off this circuit, resulting in loss of muscle strength with high amygdalar activity.

But all of this is very odd. What reason is there for any connection between strong emotion and muscle weakness? Why would our brains be designed to have these connections between the amygdala, the neurological cornerstone of emotions, and brainstem nuclei causing muscle weakness? It seems nonsensical from an evolutionary perspective to go weak at times of heightened emotion. The last thing you would want is for your legs to collapse from under you when you are terrified by a predator – and what would be the benefit of weakness with laughter?

A recent intriguing but unproven hypothesis is that cataplexy is related to a phenomenon of 'tonic immobility'. This state represents the feigning of death that many animals perform when under attack. The opossum that rolls over and plays dead when threatened is the classic example, but tonic immobility has been described in a wide variety of animals, including sharks, chickens, pigs and snakes. Although some animals remain in a particular posture, the muscles often go very slack, and other animals go floppy. While tonic immobility has not been described in humans, the phrase 'to be paralysed with fear' suggests that something similar may also occur in us.

So it has been suggested that there are similarities between tonic immobility and cataplexy, and that the connections between the amygdala and brainstem are an evolutionary hangover, but this still does not explain why positive emotions such as laughter or joy could trigger this pathway. Nor does it explain why cataplexy is more likely to occur in the presence of friends and when relaxed – it is rare to see cataplexy in the clinic, because people are usually anxious about coming into hospital to see a doctor – but it may at least provide an explanation for the existence of this circuitry, the link between emotions and paralysis.

* * *

I touch base with Adrian a few weeks after starting him on his treatment. The response is startling and rewarding. His collapses have almost completely stopped on a small dose of medication. '[The medication has] almost been like a light switch,' he tells me.

> I don't think I've had any full collapses since we started that medication. I've had what I would call a kind of near-collapse – this sort of weakness that I used to feel before a full-on collapse. I've felt at different times that I was right on the very edge of being controlled. I've had these kind of situations where I'm trying to be a little bit humorous and just kind of felt . . . It's a very difficult feeling to describe. It just feels like a sort of a swirling feeling in my lower back. But just as it has started to appear, then it's gone away.

Over the next couple of years, Adrian and I see each other regularly. His cataplexy remains well controlled, but it becomes obvious that he is becoming a little more sleepy. Initially he puts this down to his stressful job, his limited sleep, his long commute, but eventually we realise that this is not the only reason. I bring him back in for a sleep study, to see if this time we can demonstrate with more certainty that he has narcolepsy. To my surprise, once again his sleep study is totally unremarkable.

Driven by his and my curiosity to understand his condition, however, we decide to repeat the spinal tap to assess his hypocretin levels. When I see him again to review the results, his hypocretin is the same as it was two years previously. In fact, *exactly* the same! On closer inspection, it becomes clear that the lab has actually sent me the results from two years previously. I call through to hear that, actually, the new tests show that Adrian's hypocretin is almost completely undetectable, some three years after the onset of his cataplexy . . .

* * *

While the loss of hypocretin explains the features of narcolepsy with cataplexy, there remains the question as to why these hypocretin-producing neurones disappear in the first place. What has caused the vanishing into thin air of these neurones in Adrian's lateral hypothalamus? Clearly it is not as straightforward as in the Stanford dogs, where a single genetic mutation results in a failure to detect hypocretin. No single genetic abnormality has been found in humans, with the exception of one person, mentioned earlier.

It was in the 1980s, during research into a variety of diseases of unknown cause, that scientists found that most patients with narcolepsy carried a particular genetic marker. Narcoleptics carried a particular type of genetic variant of the human leucocyte antigen (HLA), a complex of proteins involved in the regulation of the immune system, first identified in the search for an explanation to tissue rejection following organ transplants. Most patients with narcolepsy with cataplexy were found to be positive for a variant called HLA DR2, a protein complex responsible for presenting fragments of infective agents – antigens – to the white blood cells that combat infection.

This was the first indication that the immune system might play a role in the development of narcolepsy; many other conditions associated with particular HLA types, like lupus and rheumatoid arthritis, have a very clear autoimmune basis. Subsequent studies have confirmed an even stronger association with a related HLA type, called DQB1*0602, present in almost all people, up to 98 per cent, with cataplexy. But this HLA type is also not the complete explanation. About one in four of the population also carries this HLA type, while narcolepsy is much rarer, affecting about one in 2,000 people. Additionally, more recent genetic studies have also demonstrated that other genes increase the risk of narcolepsy, such as those that encode receptors on T-cells, the white blood cells that are the mainstay of our immune system.

So, narcolepsy may be a type of autoimmune disease. Perhaps it is Adrian's own immune system that somehow mounted an attack on his hypocretin-producing neurones, causing this devastating loss. But, if that is the case, and this destruction is so strongly linked to the genes that influence how our immune system functions, what triggers this abnormal immune response? And why does it only occur in a tiny fraction of people with the right HLA type? A possible explanation may be forthcoming. Observations have long shown seasonal fluctuations in the onset of narcolepsy throughout the year, suggesting a link with winter infections, like influenza or streptococcal throat infections. Indeed, narcoleptics often report influenza in the year prior to onset of narcolepsy.

It was the H1N1 swine flu global epidemic in 2009–10 that gave us the strongest indication of a link between infection and narcolepsy, however. I remember that winter well. The media was full of stories regarding the aggressive nature of this new strain of influenza sweeping across the world. The NHS had stockpiled large quantities of Tamiflu, a treatment said to lessen the effects of influenza; our intensive-care unit had beds set aside for swine flu patients; and there was a general feeling of mild panic in the air. No one knew what to expect, but the omens were bad. Within a couple of months, thirty countries around the world had already reported cases of H1N1. A concerted global campaign of vaccination for the H1N1 influenza strain was set in motion, and I recall waiting in line at the hospital for my jab. Ultimately, that year's flu season passed without it becoming the public health disaster people feared, but we certainly had some very sick patients in the hospital, and some deaths.

It was within a year of the vaccination campaign, however, that researchers noticed something rather unusual. Several cases of narcolepsy appeared to be related to one of the H1N1 vaccines used in Europe, called Pandemrix. The numbers of cases of narcolepsy rose dramatically in countries where this vaccine was used, particularly in children, but a similar increase in numbers was not seen in the USA, which used a different vaccine. In later years, the numbers of new cases dropped down to levels prior to this vaccination programme. It was not only the vaccination that was associated with onset of narcolepsy, though. In China, cases of narcolepsy were found to be related to the H1N1 influenza virus itself. Interestingly, in mice bred without intact immune systems, the H1N1 virus has been shown to migrate to the hypothalamus and brainstem, causing sleep disruption directly. Expression of high levels of viral components in these areas of the brain may make them particularly vulnerable to autoimmune damage, if the immune system is intact. So, it is not just the Pandemrix vaccine that appears to trigger narcolepsy, but the flu virus strain itself.

Subsequent studies, including our own, have demonstrated this strong association between Pandemrix and narcolepsy. The vaccine has been associated with a two- to twentyfold increased risk of developing narcolepsy. But this marked increase in cases has not been seen with the other commonly used H1N1 vaccine. The reason for this is not entirely understood, although there are subtle differences between the two vaccines, in the nature and quantities of the viral fragments contained. Studies have shown that the chemical structure of fragments of the H1N1 virus are very similar to bits of the hypocretin receptor, implying that the immune system, in particular circumstances, may mount an immune response against the virus or vaccine that also erroneously targets aspects of the hypocretin system. The hypocretin-producing cells are 'collateral damage'.

This concept of collateral damage of the immune system is not a new one in the world of neurology. There are many neurological conditions that have a similar basis. Guillain–Barré syndrome, a devastating and life-threatening destruction of peripheral nerves, has long been known to be triggered by *Campylobacter*, a bacterium that causes food poisoning, and post-infection and post-vaccination damage to the brain or peripheral nerves is well recognised. Infectious agents will often use 'molecular mimicry' – appearing structurally similar to the body's own molecules – to outsmart the immune system. But if your immune system is configured in a particular way, it is liable to recognise these 'mimics', destroying the infectious agent but also aspects of your body that look chemically identical.

For narcolepsy, it appears that Pandemrix is somehow better at triggering this response than natural infection or the other vaccine widely used, if you carry the HLA DQB1*0602 type. Perhaps this is due to the process of production of the vaccine.

While the precise nature of this immune process remains to be unravelled, class actions are currently being prepared against the makers of the Pandemrix vaccine.

* * *

It is now almost six years since I first met Adrian, and his narcolepsy and cataplexy remain extremely well controlled on small doses of a medication for his cataplexy and a stimulant to help keep him awake during the day. Despite his condition, he leads a very full and active life. He has an important management role in a bank, works incredibly long hours, has an awful commute and is a father to two young children. His narcolepsy does not get in the way of what he wants to do. He appreciates the humour of cataplexy, and says: ‘If I’m honest, it was never something that really frightened me. I guess when I look back, if I was having these collapses to the extent where I was paralysed, then maybe it should have done. But it definitely started to occur in a social setting, more of a light-hearted sort of thing, and it was brief.’ He has found the journey to his diagnosis fascinating. ‘I’ve actually found this whole four or five years really interesting. I’ve really enjoyed learning about it.’

Talking to him, it is easy to underestimate the impact of narcolepsy, and cataplexy in particular. But even Adrian admits it has had some negative effects. He describes what people with cataplexy often experience, a reining in of your emotions in an effort to prevent attacks. ‘What I have become conscious of – not too much I hope – is almost avoiding that kind of situation. It depends on the scenario. At home, in the comfort of your own home, maybe you’re less controlled. But I do think subconsciously you just start to back away from the things that you know will almost without fail push you over the edge, which is a shame actually.'

For some people, though, cataplexy has life-changing consequences.

* * *

Now in his late forties, Phil is married to Kim, a few years younger, and they have two young children. He lives in a lovely part of England, and is a high-ranking manager in a large telecommunications firm. His life until recently was, to all intents and purposes, as close to perfect as could be. It is clear chatting to them both that Phil and Kim have an incredibly close relationship.

Kim says: 'We were the couple that everyone hated because we were very close. We spent most of our lives together, in line.' She reminisces: 'I don't think I have ever known Phil to have a mood, ever be unnecessarily grumpy with anyone. [He's] always ridiculously upbeat, always motivated, [the] life and soul of pretty much anything that happens in life. Top of the list to invite to a party. Always the crazy one doing very stupid things – you know, just that kind of person.' They have been together since a very young age, and obviously know each other inside out.

A couple of years ago, however, after a family skiing trip, Phil got ill with the flu. It took him a little while to recover. Some three or four weeks later, Kim and Phil were sitting on the sofa, eating dinner while watching television. Phil recalls: 'I was eating and my eyes were closing. I was almost falling asleep while eating, which is really weird, and Kim noticed it. I didn't really realise I was falling asleep, and we had a laugh about it because we thought that was pretty odd.' Kim says: 'We thought it was just a post-viral thing.'

It wasn't long before Phil noticed something else very

peculiar. 'It was very occasional. My knees would give way when I was laughing or slightly annoyed about something. Initially it was pretty mild. It was just a weird sensation. I thought it was blood pressure or something. Maybe my blood pressure was not right . . .'

I ask him when he first realised that it was something more serious. 'I remember going to the house of some friends to have some drinks one night. We were laughing quite heavily, and I remember while laughing my legs were giving way. I had to hold onto the table. I remember thinking to myself, *That is truly weird*.' A trip to the GP resulted in him being hurried into hospital. His resting heart rate was found to be low, and it was initially thought that this might be a cardiological problem.

It was the shrewd cardiologist, however, who initially raised the possibility of narcolepsy and cataplexy, and suggested a neurological opinion. I ask Kim and Phil if they were surprised to hear this potential diagnosis, but Kim had already consulted Dr Google. 'We had already looked into it and kind of established that it might be that ourselves. We googled it and came up with the diagnosis because we literally ticked every box for narcolepsy and cataplexy.'

By the time Phil got to see the neurologist, though, his symptoms had significantly worsened. I can see in his face and hear in his voice his distress at the memory.

> It was when I had the real proper full cataplexy – the fall to the floor situation – that I was absolutely certain that this is what I had. It got to the point where I was falling multiple times a day. I was unable to drive. I could not leave the house really; I could not look after the kids. All sorts of things were triggering it. I found I could not be around

> anybody that was mildly amusing. Even just greeting someone in the street, someone walking up to me saying, 'Hey, Phil, how are you?' would have me on my knees.

Kim tells me: 'It could even be a market research person just saying, "Hey, can I stop you for a second?" and Phil would just be on the floor. We were in town and we walked down to a bookshop and I just heard this slump behind me and Phil was on the floor with people all around him. It was then that we realised, actually, it was not safe for Phil to go out on his own.'

I ask Phil more about his cataplexy. 'I used to say it is a feeling like someone has just put their hand inside my head and grabbed hold of my brain and squeezed it. As a result of that, my legs would just go. I was fully conscious. I could see and hear, so I knew I had not fainted.'

In many respects he describes marked similarities with Adrian's cataplexy, but for Phil, the triggers were much more numerous. Unlike Adrian, whose cataplexy was only triggered by his own humour, Phil's triggers are not only laughter, but negative emotions as well. He tells me of one of the first major cataplectic attacks he experienced.

> The kids were messing about. I was shouting at them, and I fell to the floor, literally just collapsed to the floor. But it was not like I had been dropped. It was a slow collapse to the floor. I was lying there, and I could hear the kids laughing because they thought I was just play-acting. I remember being aware, I could see and I could hear. That was quite terrifying. Then within maybe twenty seconds, I was back up on my feet absolutely fine.

But Kim describes even more alarming cataplectic attacks.

> He would literally go down like a stone, just hit the floor. There was one occasion, when he was on the exercise machine upstairs, and the children came running downstairs shouting, 'Mummy, Daddy's fallen off the exercise machine!' Another time he got cross with one of the kids, and he collapsed over a high-back chair. He could not get off the chair because he was having a cataplectic attack over the top of it, the chair back smashing into his ribs. He was almost convulsing.

Phil shows me some family videos, of him on a trampoline or playing crazy golf with his kids, suddenly slumping down in partial paralysis. In another video, he is wielding a sledgehammer, trying to knock down a wall in his garden. As he swings back the sledgehammer, he suddenly loses strength. The sledgehammer almost drops out of his hands, and he goes limp. Other triggers would be equally strange.

> Oddly enough, a fly, just a housefly flying into the house. I would see the fly and think, *Ah, I can't stand flies, let me get rid of it*, and I would get a swat. I would go to swat the fly and I would be on the floor collapsed with a cataplectic attack. I still can't explain that one. Also hot things. I went to help a friend at a barbecue. As soon as I got near the heat I nearly fell on the barbecue, because for some reason it is perhaps a danger thing. Once, when making a pizza, I nearly fell in the oven. Also things like just getting a brush, a garden brush to brush up the leaves. If I put the brush down to brush, I would have a cataplectic attack. And playing with the kids, you know, throwing a ball, running a bath, that kind of stuff. [There were] so many triggers it was insane.

When even minor day-to-day activities or emotions trigger you to collapse, it must be hard to see how life can go on.

And it was not just the cataplexy Phil had to deal with. He describes an incident that occurred early on after the start of his narcolepsy. Phil and Kim had travelled in separate cars, Phil with the kids in the back, following behind Kim. Phil says: 'I was in the car with the kids. It was a convertible, so the roof was off the car, and I fell asleep at a traffic light. The kids woke me up saying, "Daddy, Daddy, the lights are green!" and I was like, *Oh my God, that is really odd.*' When they got home, he told Kim what had happened and she immediately took the car keys from him.

The impact of Phil's symptoms has been life-changing. He was left exhausted, unable to drive, unable to care for or play with his children, unable to leave the house alone. From a man used to being the party animal – sporty, active and successful at work – he was reduced to a shadow of his former self. 'We knew that Phil was not going to die. It was not like he had cancer or we had found something horrendous that might take his life,' Kim says. But after the practical aspects of getting a diagnosis of narcolepsy, the reality of the situation began to sink in. 'It is the illness that sucks the joy out of life,' Kim continues. 'Anything that you enjoyed, or was fun, or that brought any kind of nice stuff to your life, it just got sucked out of ours because Phil could not do it any more.'

Immediately after diagnosis, Phil was started on a stimulant to keep him more awake. Some of the older stimulants, the amphetamines, have an additional effect of improving cataplexy, but Phil commenced modafinil, a newer stimulant familiar to students as a 'mind enhancement' drug, frequently ordered online to help people study. There was no change to his cataplexy. In fact, it was getting worse.

The standard treatments for cataplexy are antidepressant drugs, which have various effects on brain chemistry, including increasing the availability of serotonin, noradrenaline and dopamine, and some also reduce acetylcholine. Their precise mechanism of action in cataplexy remains uncertain, but boosting noradrenaline increases activity in one of the nuclei in the brainstem called the locus coeruleus, fundamental in the maintenance of muscle tone.

Such was the severity of Phil's cataplexy, however, that his neurologist managed to obtain funding for a newer drug called sodium oxybate. This drug is related to GHB, a street drug and a drug abused by bodybuilders to add muscle. As a pharmaceutical agent, it is enormously expensive, and therefore hard to get hold of. An incredibly salty-tasting liquid, it is drunk twice a night, and almost acts as an anaesthetic, pushing people into very deep sleep. Sometimes that sleep is so deep that patients experience urinary incontinence, unable to wake to empty their bladders. But it is often very effective, treating poor night-time sleep, the daytime sleepiness, and the cataplexy of sufferers of narcolepsy. How it works remains unclear, though we know it binds to receptors for a neurotransmitter called GABA, and it is postulated that the locus coeruleus is also under the influence of this chemical. The theory is that sodium oxybate therefore desensitises the locus coeruleus to inputs from the amygdala, thus reducing its response to emotional triggers.

Phil's response to sodium oxybate was immediate. He remembers: 'It was instant. The night I took it, the next day I woke up and I could feel the difference. More energy. It was astonishing. I have read a lot about other people who take it and they struggle with it and they take a long time for it to work, or they feel sick. [For me, there was] nothing. Literally no side effects. It worked straight away.'

Prior to the sodium oxybate, he was falling down about ten times a day, but with the medication, he dramatically improved. 'Maybe once or twice a week I'd fall. Most of the time I wouldn't actually fall, and I was then also learning some coping mechanisms, like holding my arms or getting myself down to the floor, that kind of thing.' But he was still far from fully controlled. Reluctant to increase the dose of an incredibly potent drug, he was still experiencing significant sleepiness, and the cataplexy was ongoing, albeit much less severe.

When I met Phil, I persuaded him to increase the dose of the sodium oxybate, which was simply not high enough for him. He says of his cataplexy currently: 'It is very rare now, and if I have it, it is very mild. There is no way I fall.' He pauses for thought. 'Actually, my kids have a knack of pushing the limits just that far that I probably don't fall, but I sometimes have to go to my knees.' It seems that sometimes the exasperation and frustration of dealing with one of his children, something that any parent will be familiar with, is the one remaining trigger for Phil.

He remains sleepy, however, and there is work to be done with fully managing his symptoms. I ask them both how the condition affects their relationship, and their family life. Kim is disarmingly frank.

> I feel like I have had to get used to being married to a different person. The old Phil sneaks back in quite a lot nowadays, which is nice. But this is all dependent on whether Phil has had a good night, how he is feeling during the day, and what he does during the day. But it has massively changed how we are as a couple. We don't go out very much. Going out for a meal just does not hold

the same joy as it used to because he wants to go to sleep immediately after eating. So, going out for a meal is not a pleasurable experience.

She harks back to the days of smugness at their perfect relationship. 'That incredible relationship now is just very different. My friends all joke and say, "Ha ha, welcome to what married life is really like!"' She and Phil agree that the previously happy-go-lucky, life-and-soul-of-the-party Phil has disappeared, at least for the moment. Phil tells me: 'I would say that I am still the same person that I once was. But I frustrate myself because I still want to do the same things. But I just don't have the motivation. I don't have the strength. I come down [from the home office] just to say, "Hello, how are you doing? I'm just going to go for a nap." And then I disappear again.'

His sleepiness has changed the way he interacts with his children, too. He finds he is more short-tempered and snappy. The previous drive that Phil had has gone. Whereas in the past he would always be the one to get the family out of the house, their roles have switched. Kim says: 'He is literally the polar opposite of what Phil was before. It was always me saying: "Oh, do we really have to go out again? Can we not just stay in?" and Phil would say: "Oh, come on!" Now he just doesn't have that desire to do anything, so I find I instigate a lot.'

The sense of loss is palpable, and the uncertainty of what the future holds hangs over them. But Kim adds: 'Nothing is going to break us apart. It just means that we have to work harder as friends and be very much more mindful of each other. But it is really hard for Phil to do that, because he literally goes through his entire life exhausted.' I tell them that I am optimistic that we can improve things further.

For both of them, the attitudes of others adds insult to injury. 'It is an invisible illness. To other people, Phil looks like he is fine and he is normal,' Kim says.

> People don't understand the condition at all. We struggle with that as well. Phil, bless him, has to explain himself a million times to certain people every single time he sees them, because they still don't get it. I don't think that people appreciate how much pressure and strain it puts on you as a family unit, as a couple, as a person dealing with it. And how much effort and energy it takes to manage it all and keep going with your life, your business, your kids, your family, and your friends.

She continues:

> People still think it is really funny. We will go to an event or something like our children's performances, and afterwards someone might joke: 'Did Phil stay awake during it?' I ask myself, 'Why is that funny to you? Why is that even a question? Why do you even consider asking that? If Phil had cancer, would you ask, "Did Phil feel really ill during the performance?" You wouldn't ask it! So don't ask. It's not funny.'

I think Kim's anger stems from a desire to protect Phil. He is slightly more forgiving. 'I don't mind having a laugh about it. I think if they understand it then we can laugh about it. But if you don't understand it and you are ignorant, then I don't find it that funny.'

* * *

Looking back at my fascination with that particular page of *Neurology and Neurosurgery Illustrated*, it is perhaps unsurprising that I have ended up running a large clinic for people with narcolepsy, although my route here has been unconscious, without an active plan. In the days of being in medical school, narcolepsy was seen as a bizarre and mysterious condition, poorly understood, and was often the subject of jokes. Now, sitting in my clinic and seeing patients with this condition day in, day out, the terrible devastation that this condition has, on almost all aspects of life, is all too evident.

Studying this purest of neurological disorders gives us insight into the regulation of normal sleep, however, and the importance of several chemicals and neuronal circuits. It gives me a newfound appreciation for the ability to stay awake. And it also has some implications for people who can't sleep. Identifying the role of hypocretin in maintaining wakefulness has permitted the development of new drugs to help people sleep. Blocking hypocretin, perhaps by inducing temporary 'narcolepsy' overnight, is a potential treatment for insomnia.

For me, narcolepsy is a prime example of how a distinct but tiny lesion – an area of damage – in the nervous system gives us a window of opportunity to understand the workings of the human brain. Through an unfortunate series of factors – a genetic predisposition, a configuration of the immune system, and exposure to an environmental trigger – this disease represents an experiment of nature, and illustrates how the loss of a few brain cells can turn our experience of sleep and wake upside down.

New treatments for narcolepsy are currently being developed and some have already become available. The most obvious way to treat this disorder would be to replace the missing hypocretin. In rats, some results have been achieved by trickling hypocretin directly into the brain. This is not practical in humans, and intranasal preparations – sprayed into the nostril to be absorbed through the mucous membranes – have been trialled, but so far results have been underwhelming.

For the moment, there is no cure for narcolepsy. Researchers have tried to suppress the immune system in an effort to prevent its attack on the hypocretin-producing neurones of the hypothalamus, but it seems that by the time the symptoms of narcolepsy appear, the damage has already been done. The opportunity to intervene has passed.

Other strategies may hold more promise. Perhaps stem cell transplants, to regenerate the hypocretin-producing cells in the hypothalamus. Or gene replacement therapy, using viruses to insert active hypocretin precursor genes into remaining cells to encourage them to produce more hypocretin.

These technologies are in their infancy, but consider how far we have come: from a mysterious disease to a fundamental understanding of the causes and origin of narcolepsy over only a few decades. There are reasons to be optimistic, so watch this space.

7

BUZZING BEES

A long time before I actually get to meet David*, he sends me a detailed preparatory letter. 'Dear Dr Leschziner,' he writes, 'I understand from my partner Debra* that you have agreed to have a telephone consultation with me with regard to a serious sleep problem that I have, which unfortunately is presenting itself with more enthusiasm as time goes on.'

He then continues to give me some background, detailing his age – seventy-two – where he lives – outside Dublin – and his successful career. It is clear that David is incredibly active, both professionally and physically, cycling huge distances on a weekly basis and sailing regularly. He says he has a great relationship with his supportive partner, and his life sounds full and happy. But there is a fly in the ointment.

'Sleeping difficulties are not new to me, nor to my siblings, in that my brother, two sisters and I all have and always have had sleeping difficulties.' He describes terrible insomnia starting in his twenties. He initially put it down to the fact that his life was so full and 'I simply could not wait for the next day to come to get on with pushing my interests to new boundaries'. But his insomnia gradually worsened.

'My period of sleep is at the most two to three hours of sleep a night on a very disruptive pattern. The shortest sleeping period is sometimes ten minutes, and at the longest would never be more than an hour, and that would be a most luxurious experience.'

It is apparent that, over the years, David has seen several physicians for his sleep. At one point he was diagnosed with sleep apnoea. He has also been prescribed countless sleeping tablets. 'I have had short-acting pills, long-acting pills, Valium, Librium, antidepressants. With all these medications I had to stop taking them for the same reason, that while it might have helped me marginally to sleep, the next day would effectively be a write-off.'

The letter carries on over three sides of densely typed text, and my overriding impression is of someone who is battling furiously against their insomnia, trying desperately to carry on with their busy life, but failing to make progress. It ends with him telling me he has seen a Dublin psychiatrist, and he signs off despondently: 'I had a discussion with [the psychiatrist], and at the end I asked when the next appointment would be. He said that he did not need to see me again because he felt that this pattern was so established that there was nothing he could do about it, and that was some fifteen years ago.'

A couple of days later is our scheduled telephone consultation, me in my clinic room in London, David and Debra in their house outside Dublin. He speaks with a gentle Irish accent, and sounds vigorous, garrulous and philosophical about his insomnia. He is clearly a man of words, and uses them precisely and poetically. His main concern is that his sleep is not letting him live his life to the full. We run through the contents of the letter, but there are several complaints that he mentions, alluded to in his letter, that catch my attention.

He reiterates the very long-term difficulties in getting to sleep and staying asleep, getting out of bed approximately twenty times a night, but goes on to describe some rather unusual symptoms. One thing that he has noticed is a strange feeling of vibration in his chest that often starts when he gets into bed. As soon as he moves, this vibration ceases, but comes back soon after he lies back down. In addition, of late he has also noticed some irritability in the right leg when lying still in bed. He experiences an urge to move his leg, and if he moves it he also gets instant relief. He deals with both the chest and the leg by getting out of bed. Furthermore, for the past five years or so, Debra comments that he has been active and agitated at night. She describes him thrashing his arms and legs about and punching the headboard of the bed. It usually starts within fifteen to twenty minutes of him getting into bed and can go on for several hours. He himself is not aware of this arm-waving, even though he will sometimes hit himself.

At the end of our telephone consultation, I tell him that I am not entirely sure what the arm- and leg-waving is, although it might represent him acting out his dreams.

Undoubtedly, however, what he is describing is a condition called restless legs syndrome, and probably restless chest syndrome.

* * *

The history of medicine is littered with diagnoses that have not been formally accepted or widely recognised. Some disorders like fibromyalgia or chronic fatigue syndrome spark furious debate and controversy among doctors and sufferers alike. Others are believed to have been made up by pharma companies eager to flog their wares. Restless legs syndrome (RLS) is one of these conditions. In 2013, a regular columnist in the *British Medical Journal*, a general practitioner, reiterated this opinion. The columnist described restless legs syndrome as a 'pharma classic', that it had an 'implausible biological basis' and was based upon subjective scales, measures of patient-reported symptoms that were 'pseudo-scientifically converted to an illegitimate numerical rating'. He had never seen a patient with a primary complaint of restless legs syndrome 'in twenty years of practice'.

In fact, the year before, I had also written a review article for the very same journal on restless legs syndrome, only to be told by a retired physician from Eagle Rock, Missouri, that:

> Restless legs are a redundancy. Our legs have large muscle groups, which are designed for movement. In my experience, restless legs respond well to aerobic exercise, such as walking, running, biking, dancing, and swimming. Please let's not medicalise life with endless diagnoses that just create a pretext for more pharmaceuticals.

I metaphorically slapped my forehead in frustration at this point.

In an effort to have restless legs syndrome taken more seriously, US physicians attempted to give it a name with more gravity, feeling that the name of the condition was detrimental to its acceptance. They attempted to re-christen it Willis–Ekbom disease, after Thomas Willis, an English physician who described RLS in 1685, and Karl Ekbom, who provided one of the most authoritative descriptions of the condition in the modern medical era, in the 1940s. Unsurprisingly, this far-from-catchy handle has never really taken off. To those doubters, however, I would invite them to my clinic to meet patients like David, and Mary Rose . . .

* * *

Mary Rose is a sprightly, twinkly-eyed 85-year-old who works as an art historian, is still writing and lecturing, still travelling the world. She has more energy than I have, yet is twice my age. Her outward appearance gives no clue as to the torture she endures in her nights, and also her days. When I first met her, she was seventy-seven, but had already suffered from restless legs syndrome for many years.

'This started a very long time ago,' Mary Rose recalls, 'but of course I didn't know that it was a condition called restless legs and people just said, "Oh, you've got cramp; you must take quinine or sleep with corks in your bed."' (I later discover that corks are a traditional remedy for night cramps, although I suspect this is not supported by a rich evidence base.)

I ask Mary Rose to describe her symptoms. 'Well, it's very difficult to describe, but I think the best way is like having bees buzzing inside the skin of your legs. You can't get at them and it makes you want to scratch your legs. Also to get up if you are in bed and walk about, because it is just impossible to lie down and sleep because one's legs are twitching in this uncontrollable way.'

She describes very typical symptoms of RLS: a terrible urge to move, usually the legs, that is only relieved by moving. This urge to move is often accompanied by unpleasant sensations – tingling, buzzing, pain, an electric feeling or cramping. From a sleep perspective, the awful thing is that the symptoms of RLS get worse in the evening and night-time, making it impossible to get off to sleep. Mary Rose says: 'I began to dread sleep. I was finding more and more things to put off actually going to bed because as soon as I lay down and my head hit the pillow, then this twitching started up. I felt quite hopeless about how I was going to get through the night.' Her sensations and her leg movements were letting her sleep no more than three or four hours a night, and, even then, in short, broken bursts. These leg movements that she describes are rather characteristic.

Patients with RLS often have two types of leg movements: some are semi-voluntary, an effort to obtain relief from the torture of their sensations, and can be suppressed; but one of the hallmarks of RLS, and perhaps one of the first pieces of evidence that this is a real condition, rather than a 'made-up' one, are involuntary leg movements, kicking of the legs or twitches of the ankle, that occur deep in sleep.

The leg jerks, termed periodic limb movements of sleep, affect 80–90 per cent of RLS sufferers, and recur every five to ninety seconds for brief bouts or persistently throughout the night. Monitoring brainwave activity during the night confirms that patients are asleep during these kicks, but these periodic limb movements add insult to injury. Not only do sufferers experience the RLS symptoms that stop them drifting off to sleep, but once they do manage to doze off, the leg movements disrupt the quality of their sleep, and sometimes wake them up. The torment then continues, as once awake they are conscious of their RLS symptoms. It is not hard to see how this condition, when severe, can be impossible to live with.

'I have had complete nights without any sleep at all, but on average I would say I would get periods of sleep,' Mary Rose explains.

> If I was very tired I'd go to sleep, then I'd wake for an hour or two and then go to sleep again and sometimes get up. If it became hopeless, I used to get up and go down to the kitchen and make myself a drink and go back to bed, which sometimes soothed my brain even if it didn't soothe my legs. I can't tell what woke me. I mean it wasn't till I was awake that I then realised that my legs were stopping me sleeping.

Despite the severity of her condition, for many years Mary Rose had continued without a diagnosis and without formal treatment. Having given up on quinine and corks in the bed, she found some ways of at least lessening her symptoms. 'I got up and I rubbed my legs quite vigorously with both hands or sometimes with an unguent of some kind, to massage them. To some extent this eased it [the discomfort]. I used to do that and then get back into bed again.'

Mary Rose is now widowed, but at the time her husband must have had the patience of a saint. Imagine trying to sleep next to someone bouncing in and out of bed all night, kicking and jerking away. I'm amazed when she tells me: 'My husband was very sympathetic and always was hoping that I would find some kind of a cure, and so he never complained about my restless legs.' She later confides that very early on they decided to sleep in separate bedrooms.

After persevering for a number of years, Mary Rose found a knowledgeable GP who gave her a diagnosis of RLS, and started her on treatment, with dramatic relief. She was given a drug called levodopa, a medication more usually given to people with Parkinson's disease, which appeared to be a miracle treatment.

* * *

Like David and Mary Rose, many patients with RLS take years to get a diagnosis, which is very peculiar, especially given how common it is. Recent studies have suggested it affects roughly one in ten adults to some degree, and probably significant numbers of children. In fact, it seems that 'growing pains' in children often result in RLS in later life, and it is likely that the growing pains are simply how RLS manifests in children. In view of these statistics, it beggars belief that the GP columnist I mentioned earlier had not come across someone with this in twenty years of medical practice.

So why is it so difficult to get a diagnosis? Part of the problem is that RLS presents itself in a number of different ways. If you are lying in bed at night having trouble getting off to sleep, and you feel fidgety, it would be normal to assume that you are restless because you are having some insomnia, rather than the other way round. Frequently sufferers don't recognise the symptoms of RLS as causing their sleep issues. Equally, the sensations that people describe are incredibly varied – a sensation of pulling, jittering, worms or insects moving, tingling, itching, aching, bubbling, fidgeting, electric current sensations, tightness, and throbbing. In some people, pain may be the predominant symptom. This range results in RLS being frequently misdiagnosed as cramps, varicose veins, or nerve problems.

But, in contrast to other diagnoses, there are some clues to point to the diagnosis being RLS. The first is this association with the evening or night-time. People rarely experience this in the morning, and often there is a particular time in the evening or night that symptoms will tend to start. Patients will often tell me that they only notice their RLS when they sit down to eat dinner, or watch TV in the evening. The second pointer is the association with movement. Keeping still makes the symptoms build, and movement provides transient relief. Mary Rose finds that walking around her house at night provides some abatement of her discomfort. Similarly, David's way of dealing with his symptoms is by getting out of bed, which is clearly not conducive to a decent night's sleep.

* * *

There are of course other reasons the diagnosis may be so problematic. David's symptoms began in his chest a long time before affecting his leg. This is very unusual, but over the years I have seen several patients in whom the symptoms have not primarily affected the legs, including cases of restless arms, restless abdomens, and even a couple of cases of restless genitals. In fact, RLS, despite its name, may affect any part of the body. There have even been cases described of 'phantom RLS', symptoms occurring in an amputated limb.

When I speak to David again, shortly after his sleep study, we go through the results of his overnight admission. It is an unmitigated disaster. Over the course of the night he manages a total of only twenty minutes of sleep. There is no REM sleep, so I cannot tell him if he is acting out his dreams. For the brief period of sleep captured, it does not look like he has significant breathing issues, but twenty minutes is far from enough to refute the diagnosis of sleep apnoea he has previously been given in Dublin. Closer inspection of the overnight video, taken with an infrared camera, reveals something curious, however. The grainy black-and-white images do not provide sufficient detail for me to know what David actually looks like, but although Debra describes his arm-waving as occurring in his sleep, and David reports no memory of this activity, I see that as soon as he becomes drowsy, definitely before he is asleep, his right arm starts waving up above his head. As he does this, he wakes himself up, and the pattern repeats itself over and over again. I have seen something similar to this before. It strikes me that David's movements might represent a way of relieving RLS symptoms in his arms, and that these movements lie somewhere between voluntary and involuntary actions. I wonder if this is simply a manifestation of restless arms, in addition to his restless chest and restless right leg.

We have a slightly difficult conversation, and I try to manage expectations. I tell him that I think we need to treat what we know is there – the various manifestations of RLS – but warn him that frequently even once these symptoms are treated, the insomnia that has resulted from years of sleep disruption persists and also needs subsequent treatment. I suggest we start him on a dose of ropinirole, a treatment for Parkinson's disease, also licensed for RLS, and review his condition once he is established on treatment.

* * *

When we first meet, Mary Rose is in real trouble. The medication given to her by her GP initially worked wonderfully, almost completely abolishing her symptoms. But over the decade or so she has been on the levodopa, the RLS symptoms have become increasingly intense, and have been coming on earlier in the day. By the time she finds herself in my clinic, she is on an enormous dose of levodopa, and now also a massive dose of ropinirole, the same drug that I have started David on.

As she sits in front of me, she can barely keep still, and her legs wriggle, jerk and twitch as she tells me her story. She is totally matter-of-fact, not quite stiff upper lip, but it is clear she simply wants to describe her condition without appearing to complain. Despite her reticence, however, it is obvious that the symptoms are now unbearable, starting as soon as she wakes up in the morning, with limited relief by movement. She is stripped of the pleasure of being able to sit in an armchair and read a book. The 'buzzing bees' sensation has escalated. 'It's much worse than an itch. I can only describe it as being bitten. Not just nibbled at, but something biting you. It's really very painful.'

She is barely sleeping. She finds the tight space of a plane seat impossible, and just has to wriggle, or pace about or stretch her legs in the aisle. 'In a confined seat in the theatre, when you have to keep completely quiet, I can't concentrate on the play. I'm in agony trying to keep still.' Her garden is her main therapy; viciously attacking the weeds keeps her mobile enough to provide relief.

I am struck by the parallels between her condition and forms of training for military special forces, when recruits are thrown into a pool with their hands tied behind their back. If they stop kicking furiously, they drown. She, too, cannot stop moving. And within a first few minutes of our appointment, it is apparent that Mary Rose is a victim of a phenomenon called augmentation.

The standard drug treatment for RLS is a group of medications that increase levels of a neurotransmitter called dopamine in the brain. It has long been known that these drugs, termed dopaminergic agents, are a rapid and effective treatment for RLS, and indeed this response has been used as a diagnostic criterion, a feature to support the diagnosis of RLS rather than mimics of the condition. Levodopa acts by increasing brain dopamine levels, and other drugs like ropinirole mimic dopamine. In contrast, drugs that block dopamine, like antipsychotics, can worsen RLS dramatically. This may of course give us some insight into the causes of RLS, and certainly implicates problems with dopamine regulation in people with RLS. While this beneficial effect of dopamine-boosting on RLS has been known about for decades, there is a darker side to these drugs. Especially at higher doses, they can actually drive RLS, resulting in a dramatic worsening called augmentation. High long-term concentrations of these drugs, perhaps also due to fluctuations in levels over a 24-hour period, can cause the symptoms of RLS, the urge to move with unpleasant sensations, to come on earlier in the day, and become more intense.

Not only that, but augmentation can also result in less relief on movement, and a spread to other parts of the body, like the arms, trunk or face. Your medical salvation has become your tormentor. And when Mary Rose and I meet, she is on a dose of ropinirole nine times more than the maximum I would ever use, in addition to very high doses of levodopa, a drug no longer used in RLS in the UK because almost everyone augments on it. In the short term, every step-wise increase in her dose has led to transient benefit, only for it to drive the underlying symptoms onward in the long term. She is one of the worst cases of augmentation I have ever seen.

These drugs also have other dramatic consequences. Dopamine, as well as being a neurotransmitter influencing movement, is also fundamental to how the brain is rewarded. The pleasure associated with shopping or gambling, for example, is mediated by dopamine. But ropinirole and other drugs in this class of dopamine mimics, termed dopamine receptor agonists, can mess up this reward system and cause it to go haywire.

In recent years, we have become aware of a side effect of these drugs called impulse control disorders. Occasionally, patients on dopamine receptor agonists show striking changes in behaviour, involving activities that generate reward – things like compulsive gambling, excessive shopping, compulsive eating or hypersexuality. And patients are often not aware that their behaviour is different. It is only when they come off the drug and their behaviour normalises that they realise that a change has occurred.

First described in patients with Parkinson's disease, who take these drugs at much higher doses, we now understand that these changes in behaviour are not rare in patients with RLS either. My neurological colleague who specialises in Parkinson's disease tells stories of her patients who have gambled away huge amounts of money or taken a sudden and persistent interest in online pornography. The worst case I have seen is rather less spectacular. One man I was treating would spend about £50–100 per month on his model-car collection, but after starting him on this treatment, he spent £1,000 in the first month. Fortunately, I warn everyone and their family about this potential side effect, and the drug was stopped before his collection expanded exponentially!

Towards the end of our consultation, Mary Rose and I discuss how to improve matters. Without getting her off her medication, at least for several months, it is going to be hugely difficult to help. If we leave her medication as it is, there is a good chance that this augmentation will worsen even more. However, cutting down is going to be very problematic. In the short term, without anything to replace it, she will curse me for making her condition worse. Already sleeping only a couple of hours a night, sleep is going to be totally impossible. So there is no option but to treat her symptoms with something else, while we withdraw the medication.

* * *

So, what underlies RLS? Could it really be a made-up condition, a work of fiction by the pharmaceutical industry to flog us more unnecessary medicines? There is really no diagnostic test for RLS, not like a heart trace for a heart attack, a blood test for anaemia, or an MRI scan for a brain tumour. Perhaps it is the lack of an objective test that creates doubt in people's minds, leading them to believe that it is psychosomatic, not real. The same can be said for migraine, though, as well as some other recognised conditions.

There are certain groups of people in whom RLS is much more common. We know that RLS is more likely to occur in people with iron deficiency. Even ancient Chinese texts, attributed to the Yellow Emperor in discussion with his physician Qui Bo, describe a condition very much like RLS being treated by the administration of 'iron dust to the patient following meals'. This association was also noted by Karl-Axel Ekbom, a Swedish neurologist who provided the first 'modern' description of RLS. Donating blood can trigger the condition, and giving people iron supplements can treat it.

One of Ekbom's contemporaries, Nils Nordlander, first reported treating RLS with intravenous infusions of iron in 1953. Recent imaging studies have fairly consistently shown lower levels of iron in multiple areas of the brain, particularly the substantia nigra in sufferers of RLS. The substantia nigra, meaning 'black substance', is an area deep in the brain, dark in colour due to the cells being packed with neuromelanin, a form of the pigment that colours our skin. This neuromelanin is a precursor to dopamine, identifying these cells as dopamine-producing neurones. This lack of iron has been confirmed on postmortem analysis of the substantia nigra, and may well explain RLS even in people with normal blood iron levels. In some RLS sufferers, it appears that there may be a problem with transport of iron into the brain.

So, we think low brain iron levels are linked to RLS, but what about dopamine? We know that drugs that boost brain dopamine levels treat RLS, and that drugs lowering dopamine levels or blocking dopamine receptors can trigger or exacerbate this condition. Studies have shown that in the brains of RLS patients, turnover of dopamine is increased, and levels of dopamine are higher, but the number of dopamine receptors is reduced. At first glance, this does not appear to make sense at all. Why should a condition defined by higher than normal dopamine levels be treated with more dopamine?

Well, the current hypothesis is that in response to having too much dopamine floating around, the brain reduces the number of receptors sensitive to it. This is fine during the day, but dopamine levels are under circadian influence; higher in the daytime before dropping in the evening. It is when these dopamine levels drop in the context of reduced receptor numbers that symptoms of RLS start. And this hypothesis may also explain augmentation. Giving people large doses of dopamine-boosting drugs may cause the brain to reduce receptor numbers further, driving this imbalance as normal dopamine levels drop, meaning that symptoms worsen and start earlier in the day. The rationale for withdrawing these drugs from Mary Rose is the hope that a drop in circulating dopamine may reverse the reduction in receptors that has occurred after years of excessive stimulation by horse doses of ropinirole and levodopa.

And the link between dopamine and iron? How do we explain this? It turns out that iron is an important factor in the synthesis of dopamine. In the brain, iron and dopamine are intimately linked. But surely low iron levels should cause lower dopamine levels, if that is the case? This remains a very grey area, but studies on rats have shown that iron deficiency does indeed result in higher levels of dopamine outside the cells, and results in lower densities of dopamine receptors, perhaps influencing other molecules in these brain regions. This highly complex system remains poorly understood, and the 'dopaminergic theory' of RLS remains just that – a theory, albeit the most plausible one.

RLS is also more common in women, especially in pregnancy. This in part may relate to iron lost due to menstruation or to the developing child, but there is more to it than this. Women suffering in pregnancy often get an extraordinary improvement in their RLS after a week or so following the birth, which simply cannot be explained by a sudden rise in iron. There must be a hormonal effect as well. Unfortunately, women getting RLS in pregnancy are at higher risk of developing it later on in life. Kidney problems can also give rise to this condition, as can various neurological disorders.

Perhaps the most important risk factor is having a family history of RLS. Roughly half of people with RLS have other members of the family with it too, and identical twins are more likely to have RLS compared to non-identical twins, strongly implying that there are genetic rather than environmental factors at play. David's extremely strong family history of similar sleep problems may well represent a strong genetic predisposition to RLS. Huge international efforts to identify genes that contribute to RLS have been undertaken. At the latest count, these have involved 45,000 patients with RLS across multiple countries, and have identified nineteen genes that appear to raise the risk of developing this disorder. These genes are all implicated in the development of the nervous system, influencing the growth of neurones, neural circuit formation and how synapses – the communication points between nerve cells – are made.

But knowing which genes are involved, and which neurotransmitters are at play, does not give us the ultimate explanation for what causes RLS, only a few tantalising clues and further avenues of research down which to proceed. However, knowing that there are genetic contributions, understanding some of the chemical changes in the brain and being able to demonstrate periodic limb movements in sleep on an overnight study should at least persuade my more cynical colleagues that this is a condition that *does* exist, that is not a figment of the imagination or a mere marketing ploy.

* * *

I first meet David and Debra face to face in clinic a short while after the sleep study. Neither of them is at all what I expected. David is immaculately dressed, looks in robust health, and shows no outward evidence of a lifetime of poor sleep. Having chatted to Debra on the phone as well and having heard her accent, I was expecting a sixty- or seventy-something-year-old Irish woman, but she is youthful and exotic-looking.

Grateful to finally meet them in person, my heart sinks as David tells me he has only managed to take two doses of the ropinirole before stopping it. Despite an extremely low dose, he tells me that on both occasions he felt rather down on waking up after taking it, and so has not persevered. It seems that our relationship is doomed. After chatting to him further, however, I am more and more convinced that what he is describing is indeed restless legs syndrome, restless chest syndrome and restless arms syndrome. We decide to give it another shot, but this time with a similar drug that is administered by skin patch. The dopamine receptor agonist trickles in at a constant dose, absorbed through the skin, for twenty-four hours a day. Perhaps by eliminating fluctuations in the drug level, we can prevent further problems. I cross my fingers and hope.

Within a few days, my heart sinks as I see an email from Debra in my inbox. On opening it, though, I find a rather joyous message from her. 'David started on the 1mg patch,' she writes,

> and that evening he slept solidly bar one or two visits to the bathroom for almost fourteen hours . . . Miracle! Second night and again a solid night's sleep but a slight grogginess the following morning. Third night, he decided not to replace the patch at the usual time, and after initially falling asleep, he woke slightly agitated within his

upper body (about thirty hours using the same patch), so he replaced it and had another successful night's sleep. I know it's early days, but having tried so many things, we are both pinching ourselves with anticipation that we've finally found a solution.

I breathe a sigh of relief in the knowledge that we have made a breakthrough. The clear improvement of the vibrating chest sensation with the medication, and the return when the patch runs out, really cements the diagnosis of restless chest syndrome in my own mind. I email back, saying that it is indeed early days, and the grogginess after only a few days of treatment may simply be David paying back a sleep debt accrued over decades, but we need to see if this resolves.

A few weeks later, another email from Debra:

'Just writing to say that [the treatment] has transformed David's life. Such quality sleeping – it is a miracle. Long may it last!! It must be six weeks now of undisturbed sleep and his quality of life has so much improved. David has spent almost his entire life trying to manage without sleep. He is like a new man.'

David's treatment has not been without hurdles. The skin patches have caused some skin irritation, but he is reluctant to try anything else. The smallest possible dose has had life-transforming effects. Having lived a life overshadowed by terrible insomnia, he feels better now than he has felt since his twenties, and so prefers to use mild steroid cream to treat the irritation. The patch has stopped his chest and leg sensations, and Debra reports no further arm-waving in the night. I briefly try an alternative non-dopamine medication, but David soon reverts to the patch.

At our most recent meeting, some three years after we were first in contact, David's life remains spectacularly different to what it was. He is still sleeping well, and is doing everything he wants to do without the prospect of a night of awful sleep ahead. He remains on a low dose of the skin patch, and I am keen to keep it so. Given the severity of his symptoms, he would not thank me if he were to augment in a few years' time, and so I have added in a small amount of an alternative medication which does not have the same issues.

Many of the alternatives have their own problems, however – drugs used in the treatment of RLS include certain anti-epileptic drugs, and opiates, painkillers related to codeine, all of which raise concerns of dependency. Thankfully, the evidence of dependency in patients with RLS is minimal, and my US colleagues will even use methadone, a drug routinely used for heroin withdrawal, for their most afflicted patients.

There may also be other benefits to David apart from simply getting a better night's sleep. There is a growing body of evidence that RLS, like sleep apnoea, is associated with an increased risk of cardiovascular disease, high blood pressure and stroke. When we record the limb movements of sleep associated with RLS, we find an increase in blood pressure and heart rate that is independent of whether these leg movements disrupt sleep. So, it may be that these frequent leg movements prime the cardiovascular system to age more rapidly. This area, however, remains poorly understood, and at present we do not know if the treatment of RLS or these limb movements reduces cardiovascular risk.

And what about Mary Rose? What has happened to her RLS? Withdrawal of the levodopa and ropinirole to reverse some of the augmentation has not proved easy, but has been managed with alternative agents. After a period of a year without any dopamine-boosting drugs, she has been restarted back on a small dose of ropinirole, in combination with several other drugs. Eight years down the line, her symptoms are reasonably controlled. Now in her mid-eighties, she is still managing to keep busy, still flying around the world, and still active academically. Gardening remains her own way of dealing with her symptoms.

'It's a fantastic difference in the fact that I have peace of mind,' she tells me.

> Because I know that as long as I take the pills – and there are times when I forget and I know quite soon after I should have taken them, I know because the bees start buzzing in my legs – but as long as I do what we have worked out together, I'm free of restless legs. Sometimes I get an attack which is just so awful that I find I'm walking about all night, but it's my fault because I've forgotten to take the pills, and then it comes back in full force and there's nothing I can do about it except make up the pills that I've lost and walk about until it goes.

I am not totally convinced that she is free of her symptoms. As she sits in front of me, her chair constantly creaks as she shifts and moves her legs, rotating her ankles and stretching her calves. When I point this out to her, she laughs. 'You're the only person who notices that! I didn't know that I was always moving. I was completely unaware of it!'

Her sleep remains an issue. Even in the absence of the RLS symptoms, her sleep is not perfect. Decades of disruption have had their toll. 'I'm sorry to say that I think the patterns set when I was middle-aged, now that I'm much older, have remained. They were perhaps set when I had babies and one had to get up in the night at three o'clock in the morning,' she muses. 'Three o'clock in the morning seems to be a time when I wake anyway.'

Once awake, it is her mind rather than the bees in her legs that buzzes with thoughts, and it takes her a while to drift off back to sleep. She has worked hard to avoid actively trying to sleep, which often makes the problem worse. 'I make it into a passive process by listening to my audio books or music, by which time my brain is beginning to stop running and so I do actually then feel ready for sleep.' But she is thankful at least not to be tortured by her restless legs.

* * *

David and Mary Rose are two of my most striking patients with RLS. It's worth stressing that for the majority of people with RLS, medication is not necessary. Given how common RLS is, this is a huge relief. The decision to start drug treatment is not straightforward, a choice between the devil that is the RLS and the deep blue sea of potential side effects of treatment.

Instead, minor lifestyle changes can often suffice. A reduction in caffeine and the avoidance of tobacco or alcohol can often have a major effect. Sleep deprivation also worsens RLS, and so a regular sleep pattern can help. A common situation I come across, however, is people who have been given sedatives or antihistamines for their 'insomnia', which is actually being caused by RLS. Sedating antidepressants, antihistamines and a host of other medications may worsen or trigger RLS, and so a simple switching of medication can have dramatic results. Looking for and treating underlying conditions like iron deficiency can also resolve symptoms. The single most important facet to managing RLS, though, is recognising its presence in the first place – and convincing some doctors of its existence.

Unfortunately, while we understand something of the genetic basis of RLS, its predisposing factors, and something about what happens on a chemical level in the brain, the underlying cause remains a mystery. It is proposed that whatever is going on in the brain ultimately influences neurological function in the spinal cord. Perhaps fibres projecting down from the brain, using dopamine as their neurotransmitter of choice, dampen down circuits that process sensations. When these descending projections underperform, and dopamine levels drop in the evening, the brakes on these sensory processing circuits are taken off, generating these unpleasant sensations spontaneously. This concept of the brakes being taken off may also make other regions of the spinal cord more active than they should be, and perhaps this releases primitive reflexes that are otherwise suppressed, causing the periodic limb movements in sleep that accompany RLS.

I recall being at a big sleep conference in Minneapolis a few years ago, at a lecture on sleep in animals. A video of a sea-lion sleeping while floating in the water came up on the screen. I was surprised to see its flippers periodically finning, essentially treading water, and was struck by how similar to the leg movements seen in the sleep laboratory they appeared.

Perhaps, and this is entirely speculation on my part, as with many areas in neurology and sleep, these periodic limb movements are a throwback to our ancestors. In normal life, these reflexes are dampened down, but when things go wrong in our brains, or indeed our spinal cords, it may just be that our evolutionary past comes back to haunt us.

8

SEIZED BY THE THROAT

Imagine you are tasked with mapping the floor of the world's oceans, from the shallows of the beaches to the depths of the Marianas Trench. It would be reasonable to expect access to the latest gadgetry – fleets of ships equipped with sonar, submersibles and satellite imagery. Instead, you are handed a snorkel and mask. As you wade into the English Channel to start the job, you lower your head in the waves and catch the occasional glimpse of your own hand in the murky waters. You can just about make out the muddy bottom a few feet away from your face. The rest is just nothingness, as the sea floor falls away into the depths.

When we routinely study the neurological system, similarly, we can only peer into the shallows of the brain. The depths often remain invisible, obscured.

In the world of neurology and sleep, we rely on the scalp EEG, the electroencephalogram. By analysing the electrical signals of the brain, through the use of wires stuck with glue to the scalp, we look for fluctuations in brainwave activity. The rate and the amplitude of these fluctuations help us define the stages of sleep, as do other features. And when we look for brain disease or epilepsy, we look for abnormal patterns – excessive slowing, sharpening of the patterns or spikes of electrical activity.

Yet the EEG, first used in the 1920s on humans, is our

equivalent of the snorkel and mask. These scalp electrodes are attempting to record brain activity through skin, fat, the skull, and cerebrospinal fluid. The strength of these signals is tiny compared to that put out by the simple act of blinking, the triggering of small muscles in the face and head. Furthermore, the deflections in those squiggly lines – originally plotted by pen on paper, nowadays on a computer screen – do not result from the electrical changes in a single neurone. Rather, it requires the accumulation of impulses synchronised between thousands or millions of neurones, more than 6 or 7 square centimetres of cerebral cortex – the thin layer of cells lining the surface of the brain – all orientated in the same axis. So the detail the EEG gives us is incredibly limited. Only large changes in huge numbers of neurones, all pointing in a similar direction, on or near the surface of the brain, are detectable. This technique, the staple of our investigational tool box and upon which we rely so heavily, tells us next to nothing about what is going on in the vast depths of the brain. Forget about snorkelling in the English Channel, it's more like wading through a vast swamp.

* * *

Some of the people I see in my clinic are like old friends. Janice is one of these. I have known her now for almost a decade. In my mind's eye, she is always smiling, a dazzling broad smile set in a gentle olive face, completely unlined by the passage of that decade, or indeed the fifty-something years of her life. But behind her cheery demeanour lies a difficult and troubled past. Janice is very open about her traumatic childhood. She was born in the UK and is one of seven children. Her parents had emigrated from Trinidad shortly before her birth.

'My mum's side was French and Anglo-Indian, and my father was Indian and black. My great-grandfather came from Scotland and that's how we got the family name,' she tells me.

Janice remembers her family home as being very chaotic. Her mother had mental health issues, and she recalls domestic abuse. Her pleasant memories are of her uncle taking her and one of her siblings to Hyde Park for days out, and she still associates this green expanse in central London with a sense of peace, of being carefree. She and her sister would often run away from home, to sleep on the benches in the park. 'We would bunk the buses, or walk. We didn't actually always make it to Hyde Park; sometimes we would sleep in Euston station. My sister and I would cuddle up together. The police would come along and find us.'

At about the age of eleven or twelve, Janice and some of her siblings were taken into council care, and were placed in a children's home. She has some recollection of a few white couples wanting to adopt her: 'They [the council] refused because they were white, and they said, "You can't have children like that, it is impossible,"' she recalls.

Going into care did not improve her lot in life at all. 'In the [children's] home, at my [family] home, my whole life was just filled with abuse and violence. And I wasn't well – [I was] a very sickly child. I was in and out of the hospital. I was very thin, couldn't eat properly. I was malnourished. Yet everyone ignored me. They chose just to step over me.'

It is not at all surprising that Janice had behavioural problems. She speaks of being uncontrollable, of responding to violence with violence. 'If you pushed and shoved me, you were going to get ten times back what you just gave me, which was not very nice.' Yet the home's way of dealing with her outbursts was medication. 'They drugged me up to sedate me, to make me behave. They couldn't cope with me.'

For years she was given sedatives and antipsychotic drugs, and remembers being seen at the Bethlem Hospital, the original Bedlam, by a psychiatrist. 'I didn't want to take it [the medicine] but I was forced – a child being held down, all these people surrounding you, and you are being forced to take it. I was frightened, so I started fighting with them. I said, "I am not taking the medication any more." They called the doctor one night to give me an injection.'

It was in the maelstrom of this traumatic childhood that her night-time events started.

* * *

The first time I meet Janice, she is in her late forties. In the consultation, she tells me of her past, but the full details only come to light in the months afterwards. She has been referred to me via one of my astute respiratory sleep colleagues, who in turn has had her referred to him from another sleep clinic in London.

She describes terrifying nocturnal events that have been plaguing her since she was a teenager. Over the past two or three years, however, these have been getting much worse.

I ask her to describe a typical attack. She tells me that, as she is dozing off, she feels that her heart has slowed, almost stopped. 'As soon as I drop off to sleep,' she says, she suddenly feels a sensation of being crushed, or someone strangling her. The strong feeling of being choked is terrifying, and she fights desperately for breath for several seconds before it eases. When it's really bad, she will sometimes bite her tongue. Needless to say, she is getting very little sleep. These attacks will only happen while asleep, never while awake, and are happening almost every night. 'At their worst, they will go on through the night, sometimes fifty or 100 times,' she tells me.

Her symptoms are those of sleep apnoea – collapse of the airway in sleep – but this has already been ruled out by a sleep study performed in the referring centre, and repeated by my colleague. There are some alarm bells ringing in my mind as we discuss her symptoms further. She goes on to tell me that she is sometimes aware of her right leg jerking while she is choking, just a few movements. And there are other unusual symptoms. 'I feel that maybe my tongue has got bigger during attacks, and I am actually choking on my tongue. And my mouth fills with saliva, sometimes with blood in it from where I have bitten my tongue.' She also tells me that her attacks are much worse in the few days before her periods.

Apart from the fear that these attacks cause, she feels destroyed by the terrible sleep deprivation. 'I have to go to work and just try to function as a normal human being. But I have to struggle and struggle. The only thing that keeps me going is the children.' She now works as a carer of children with special needs, in hindsight understandable given her own childhood. When she gets home, however, there is nothing left. All she wants to do is sleep, and she will often drift off on the sofa. But even then her attacks will wake her up again. She is fearful of sleep itself.

I ask her how long these events have been going on. 'I can specifically remember them since going to secondary school,' she says. When I ask why it has taken her this long to seek help, she tells me: 'I would tell my parents about these attacks, and they just totally ignored me. In my household, they had this rule: too bad if you are ill, get out and just function properly; don't make a fuss.' Later, in the children's home, 'I remember specifically telling them, "I am having difficulty every night with my breathing," and they just put it down to the fact that this is a troublesome person who has had a difficult upbringing and this is the way she behaves. "She just needs to be sedated and that will take care of it."'

Janice recalls the first time someone took her symptoms seriously. In her twenties, while staying with her sister, she had one of her attacks. Her sister was so alarmed that she called an ambulance. 'She actually believed there was something wrong with me,' Janice says. But, in the hospital, she remembers the doctors dismissing her symptoms. 'They said it couldn't have been that bad because I looked perfectly okay.' So she continued to live with her attacks. 'I did go to the doctor's occasionally and mentioned my attacks, and they said, "Oh, it's asthma," and I was just put on inhalers.'

In the absence of sleep apnoea, another explanation for night-time choking is reflux – acid coming up from the stomach and irritating the throat, causing the muscles in the throat to go into spasm. But as I listen to her story, it does not sound like reflux or asthma at all. Nor does it sound like sleep apnoea. I tell her it sounds very much like epilepsy.

* * *

Say the word 'epilepsy' and it conjures up images of dramatic convulsions, shaking of the entire body, writhing around on the floor with a purple face, frothing at the mouth, and urinary incontinence. For some patients with epilepsy, this is an accurate description.

In the brain, electrical activity is a highly regulated and precise phenomenon, and it is that precision, that subtle interaction between various neurones, that underlies all our neurological functions: our speech, vision, comprehension, movements, consciousness – everything.

But epilepsy represents the loss of control of these electrical impulses. Whether due to a genetic condition, or some irritation to the brain like a tumour, infection or stroke, if those processes that tightly control these impulses are diminished, then areas of the cerebral cortex, the grey matter that lines the surface of our brains like the outer shell of a walnut, fire in an uncontrollable manner. The neurones become synchronous, all firing simultaneously (unlike normal service, when they all talk to each other in a highly organised way), disrupting the usual activity of the brain.

Think of the deck of a ship, with hundreds of people walking this way and that, all in different directions. If all of them run together, from port to starboard and back, the ship begins to sway, and may ultimately capsize. It is these large areas of synchronous electrical activity spreading through the whole brain that result in these convulsions, activating all muscle groups, losing control of movements and of bladder function, losing consciousness.

However, for reasons that we do not fully understand, in some people seizures do not spread to the whole brain. They originate in one part of the brain, and are limited. They do not propagate widely, and the area of abnormal activity remains restricted to one or a few parts of the brain. In these cases, what these seizures – termed focal seizures – result in is not a generalised convulsion. Rather, the internal and external manifestations of these seizures are a function of which part of the brain is involved.

And not all seizures result in shaking. The commonest type of focal seizure affects the temporal lobe, in which areas involved in autobiographical memory, language, smell and emotion reside. Patients with temporal lobe seizures may experience a sudden strong smell, disruption of speech, or an impending feeling of doom. Involvement of the memory areas can give rise to *déjà vu*, that sudden feeling of familiarity that we all occasionally experience, or the opposite – *jamais vu* – where what should be a familiar environment seems novel, never seen before. If the seizure spreads more widely, it may result in jerking by involving motor areas, or result in confusion, or loss of awareness, but not loss of consciousness.

I vividly remember as a junior doctor seeing an elderly lady in the emergency department who had been found wandering the streets. This was happening frequently and her GP had raised the diagnosis of Alzheimer's disease. Upon examining her, however, in addition to confusion, she had some rhythmic jerking of her right hand, a subtle clue of ongoing seizures. An intravenous injection of an anti-epileptic drug suddenly brought her back, and both the twitching and her confusion were resolved.

Over the years, I have seen some weird and wonderful manifestations of epilepsy, like the young man who would suddenly feel he was upside down in the world, with his vision turned through 180 degrees. His seizures were affecting the parietal lobe, where our brains represent where we are in relation to the world. Then there was the sixty-year-old man whose temporal lobe seizures would result in a feeling of religious ecstasy so intense that he would feel connected to God. He refused to treat his epilepsy, as he worried that he would lose these spiritual experiences. I have seen several patients with visual hallucinations as a result of seizures affecting the occipital lobes, where the visual cortex sits.

By identifying which parts of the brain are involved in the seizure, and correlating them to the symptoms, we learn something of how the brain is organised, and how neurological functions are 'localised'. In fact, much of what we know about the localisation of functions in the cerebral cortex comes from the artificial triggering of small seizures.

In the 1940s and '50s, Wilder Penfield, an American neurosurgeon, would operate on patients with epilepsy. Before removing diseased brain, he would painstakingly stimulate areas of the cerebral cortex to make sure he was not taking an important bit of brain tissue away. Working on awake patients under local anaesthesia, he would use his electrical probe to trigger small areas of the cortex and see what the patient reported or if they moved. Through this technique, he generated detailed maps of how the brain represents sensation and movement, but also showed that stimulation of the temporal and parietal lobes resulted in symptoms similar to those described by patients having epileptic seizures in those areas – *déjà vu*, fear, memories, visual hallucinations. A variant of this technique is still used today in people undergoing certain types of epilepsy surgery.

Indeed, there is a rare natural mimic of seizures being triggered by stimulation of the cortex. Sometimes even stimulating areas of the cortex through mental activity or external stimuli can generate seizures. The commonest form of this is flashing lights causing seizures arising in the visual cortex, but, rarely, even listening to one particular type of music, writing, solving puzzles or the sensation of hot water being thrown over the head and body can precipitate what are known as reflex seizures.

So what is it about Janice's events that suggest epilepsy to me? The first feature is that they are always very similar, if not identical. Focal seizures arise in one particular part of the brain, an area of irritation or abnormal function. So, while different seizures in one person may propagate more or less widely within the brain, the onset of their seizures is identical. She defines her milder events as being differentiated from her more severe events by the lack of tongue-biting, but essentially they are all characterised by the same symptoms. If her seizures were to spread more widely, then there may be confusion or other symptoms, and if the whole brain were to be involved she may also have a full-blown convulsion (she never has). The events that she describes are highly 'stereotyped'.

And then there is this curious worsening in association with her periods. It is well-recognised that some women have dramatic changes in their epilepsy as they progress through their menstrual cycle. The sex hormones oestrogen and progesterone have profound effects on the brain, with oestrogen generally making seizures more likely to occur, and progesterone having a protective effect. In the days before a period, the ratio of oestrogen to progesterone is highest. Many women find that this is the time of the month that they are most at risk of seizures. In extreme circumstances, the combined oral contraceptive pill is run through continuously for three months at a time, to limit the number of windows during which women are at a higher risk of seizures.

Another unusual feature of Janice's attacks is that they only occur while asleep. She has never experienced them during the day, unless she has had a nap. Surely seizures would arise in the daytime too? Actually, not necessarily. Sleep and epilepsy have been known to be closely linked for well over a century. Many people report that sleep deprivation is a potent trigger for their seizures, that their convulsions will only happen or be more likely to occur after a very late night or an early morning. We utilise this in clinical practice when trying to diagnose epilepsy, by sleep-depriving patients prior to an EEG. In addition to sleep deprivation triggering seizures, it also appears to bring out the electrical fingerprint of a predisposition to seizures on the EEG. And sleep disruption due to other problems like sleep apnoea can also exacerbate epilepsy.

It is not only lack of or disrupted sleep that does this, however. Sleep itself does something to the abnormal electrical impulses that underlie epilepsy. When we bring patients into the sleep lab, we will start recording their brainwaves before they go to sleep and continue throughout the night. We will often see an entirely normal EEG pattern while they are awake suddenly transform into a very active abnormal EEG as soon as the person drifts off to sleep, even sometimes as soon as they become drowsy. The process of going to sleep in itself seems to facilitate epilepsy. One possible explanation is that, as we enter into non-REM sleep, neurones in the cortex become more synchronised, thus contributing to the uncontrolled synchronous discharges of large areas of neurones that result in seizures. In fact, in REM sleep, when the EEG is most like that of full wakefulness, and the synchronisation of neuronal discharges is least, seizures are least likely to arise, and EEG abnormalities are least likely to occur.

But that is not all. For some people and some types of seizure, it is the transition from one stage of sleep to another that triggers seizures. I have seen several patients who will have seizures when they move from deeper sleep to light sleep as a result of snoring or other stimuli. Why this is, however, I'm afraid I have no idea.

There is one type of epilepsy, though, that is most strongly linked with attacks from sleep. Seizures arising from the frontal lobes, the areas of the brain directly behind the forehead and above the eyes, are very frequently much more likely to occur from sleep, and in some patients occur exclusively from sleep. These seizures are sometimes caused by a genetic mutation in genes that contribute to ion channels, proteins that transport salts across the membrane of neurones. These genes, and the epilepsy, are passed from one generation to the next, with over 100 large families throughout the world described in the medical literature. For the majority of those with this condition, however, it is sporadic rather than genetic, often without obvious cause, but occasionally associated with a structural problem in the frontal lobes.

Typically, frontal lobe epilepsy will start in the teens, with frequent seizures throughout the night. And, as with other forms of focal epilepsy, the manifestations of frontal lobe epilepsy mirror what we know about the function of the frontal lobe. In addition to its role in planning and behaviour, the frontal lobe is intimately involved in movement. At the part of the frontal lobe closest to the back of the head sits the primary motor cortex, and seizures arising here will result in simple twitching or shaking of a body part. But further forward, closer to the face, there are areas of the brain that are responsible for regulating more complex movements, often coordinating actions involving both sides of the body, and areas controlling generation of speech. Seizures arising here will result in unusual, sometimes extremely bizarre, movements.

The kinds of activities we will see in the sleep lab include patients suddenly waking from sleep, with their legs furiously cycling in the air, arms windmilling aggressively, often accompanied by shouting or screaming. One young woman I look after sits up in bed, her arms flailing, body rocking back and forth, as if possessed by a demon. I have also seen videos of patients uncontrollably performing forward rolls in the bed, or jumping out of bed, hopping up and down with their arms raised and fists clenched, exactly like a boxer sparring in the gym. All of this activity is usually done in full consciousness, but entirely without control, the seizures themselves waking someone out of sleep.

There is also sometimes a degree of overlap between these frontal lobe seizures and non-REM parasomnias, events like sleepwalking, sleep-talking and night terrors. Some of the behaviours seen in frontal lobe epilepsy are almost indistinguishable from sleepwalking or related disorders. This may be related to non-REM parasomnias being triggered by small epileptic seizures, but it is also possible that there are inborn patterns of behaviour critical for survival, like running or fighting, coded for deep in the frontal lobes, triggered both by seizures and non-REM parasomnias. In practice, distinguishing these two conditions can in certain circumstances be extremely difficult. Sit several 'experts' around a video of a patient, and they will often have entirely different views as to what they are witnessing.

Janice does not describe any of the classic features of epilepsy, though, apart from the fact that her events are very similar each time. The confusion, the speech disturbance, the *déjà vu*, the hallucinations of smell – the characteristics of temporal lobe epilepsy – are just not there. And the fear of these events is understandable as a response to suddenly being choked out of sleep rather than as an epileptic phenomenon. Nor do her seizures sound like frontal lobe epilepsy, with the exception perhaps of the fact that her events solely arise from sleep. So, if her events do indeed represent epilepsy, where in the brain are these seizures coming from?

The jerking of the right leg in some of her events points to it coming from the left side of her brain. If these are indeed seizures, this suggests involvement of motor areas, and the left side of the brain controls the right side of the body. Her most prominent symptom is the sensation of choking, or her throat constricting and suddenly feeling throttled. And there is a rare form of epilepsy that can cause this.

Deep within the brain lies an area called the insular cortex. On either side, roughly just above the ears, the insula sits, covered by an expanse of temporal lobe from below, and frontal and parietal lobes from above, like the teeth obscured by the upper and lower lips. The insula connects to all these regions as well as the limbic system, the areas of the brain involved in emotion. And because of its location, seizures in the insula can mimic other forms of epilepsy, depending on where the seizure activity spreads. Involvement of the limbic system may result in anxiety, panic or fear, and spread to the frontal lobe can cause movements identical to those seen in frontal lobe seizures, like the kicking, cycling or rocking described above. Temporal lobe spread can rarely result in the triggering of areas involved in hearing, causing auditory hallucination like whistling. And involvement of the autonomic system, the aspect of the nervous system involved in maintenance of blood pressure, heart rate and gut movement, can cause churning of the stomach, goosebumps or, in extreme cases, abnormalities of the heart rhythm, to the point of briefly stopping the heart altogether.

The most common manifestation of insular seizures, however, is usually related to spread to the sensory cortex, the bit of the parietal lobe that overlies the insula. Looking at how sensation is organised in the parietal lobe, imagine a picture of your own body mapped across the brain. It is not true to scale, as areas of the body very sensitive to touch are better represented than those less sensitive parts. The leg, abdomen and trunk are relatively small, while the hand, face, eyes and tongue are grossly distorted, like a caricature. The area of the sensory cortex representing the leg wraps the parietal lobe deep into the sagittal sulcus, in the midline of the brain, and as we move to the side, then the trunk, the arm, the hand.

Finally, in the area closest to the ear, an area known as the operculum, the bit of the parietal lobe smothering the insula: the lips, the tongue, the throat. And spread of seizure activity from the insula to this area of the brain results in tingling or other forms of sensory disturbance in the lips, gums, tongue or throat, often associated with a feeling of choking, constriction or suffocation. Exactly like Janice. And while our understanding of insular epilepsy has significantly advanced over recent years, even in the 1950s Wilder Penfield, the neurosurgeon who mapped out functions of the cortex using his electrical probe, wrote of insular seizures: 'A sensation . . . rises to the throat . . . The sensation may be sickening or pressing and may end in a feeling of choking.'

* * *

During our initial appointment, I tell Janice that I want to look for evidence of epilepsy. She has already spent the night in the sleep lab, having seen my respiratory colleague. The sleep study only shows generally poor sleep. Sod's law, she has not had any events overnight, perhaps because this was not done in the week before her period. I organise an MRI of the brain, to look for any abnormalities that might give rise to epilepsy, but it is normal. We undertake a routine EEG, recording for about half an hour while Janice is reclined on a couch. This is also essentially normal. I organise another, this time as a sleep-deprived recording, to see if we can monitor her brainwaves as she drifts off to sleep. This time we see some slightly abnormal activity over the left temporal lobe, not specific to epilepsy, but at least suggesting that there is some abnormality of brain function in this region. I book in a repeat admission to record a whole night, to see if we can capture one of her events but, as luck would have it, again we are unsuccessful.

I am enormously frustrated by my inability to prove my diagnosis of epilepsy, and Janice is too. Ultimately, she does not care what is causing her events, she just wants them treated. In desperation, I request a PET scan. This type of study involves the injection of glucose labelled with a radioactive marker. The brain is scanned for radioactivity, to look at patterns of utilisation of the injected glucose. Sometimes, areas disrupted by frequent seizures exhibit abnormal function by taking up less glucose than they should. When I see her again in clinic following her scan, the results are definitive. On the right side, the insula glows purple and pink, showing normal uptake of the radioactive glucose. But on the left, it is obviously blue and green, the cooler colours showing much less activity than would be expected. To Janice's and my relief, the diagnosis of insular epilepsy is confirmed. We can start treatment.

* * *

The EEG is the standard way of diagnosing epilepsy. Even recording someone's brainwaves between seizures can often show tell-tale signs that certain parts of the brain are liable to generate seizures. For certain types of epilepsy, this is an extremely useful diagnostic test, but it does not always give an answer. The definitive test is to capture one of these seizures with the EEG leads attached. We see characteristic changes in the brainwaves as the seizure starts and then spreads through the brain, the signature of epilepsy as it happens. But there are limitations to the EEG; it is not foolproof.

For certain types of epilepsy, like genetic forms, or temporal lobe epilepsy, most cases can be proven simply by doing one or two standard EEGs, even between seizures. But, as previously discussed, identifying features of epilepsy on the EEG, particularly when the EEG is not recording during an actual seizure, depends upon the location of the source of the seizures. If the abnormal area is small, very deep, or even orientated the wrong way, then simply placing electrodes on the scalp may not detect any abnormalities. And this is a particular problem for seizures arising from sleep. The wrinkles or crevices that cover the brain are especially deep in the frontal lobes, and some of the cerebral cortex that lines the outside of the frontal lobes is some distance from the scalp, with bits of the frontal lobe sitting closer to the eyeballs than the surface of the brain. In frontal lobe epilepsy, the EEG between seizures is often normal. Even if actual seizures are captured, in about half of all patients the brainwaves are either entirely normal, or there is so much muscle activity related to the seizure itself that the activity of the brain is totally obliterated.

And there is the further complication of these seizures arising from sleep, so that patients and their partners are often unable to give a full description of their events. The diagnosis of frontal lobe epilepsy can be so problematic that some forms were until recent years described as 'nocturnal paroxysmal dystonia', a movement disorder rather than epilepsy. It is only with the advent of EEG recorded from electrodes implanted directly into the brain that this condition has been shown to be epileptic in origin. And in insular epilepsy, the same issue applies. The insular cortex is very deep, covered by a thick layer of other areas of the brain, distant from the scalp, so the EEG can be entirely normal as well.

Even now, I and my colleagues regularly debate whether or not the behaviours we are seeing are nocturnal epilepsy, non-REM parasomnia or some other form of sleep disorder. You may ask why it matters. Why don't we just start people on anti-epileptic drugs? In Janice's case, with the benefit of hindsight and experience, I perhaps nowadays would not have performed the PET scan. On the basis of her description and the slightly abnormal EEG, I would now likely start her on anti-epileptic drugs anyway, and have the courage of my convictions. But about a third of patients with frontal lobe epilepsy do not respond to the most effective anti-epileptic drug in this condition, and so ongoing events despite treatment do not rule out epilepsy.

* * *

Janice's response to anti-epileptic medication has been dramatic. When I see her a few months after starting treatment, she tells me that she has had some side effects, although they have largely settled. But importantly, for the first time in decades, she is having good-quality sleep. Her events are still ongoing, but are much less frequent, and much less intense. Over the next year or so, we gradually increase her dose, to the point where she only gets a few seizures in the week before her period, or when she is unwell with infections. Her transformation is amazing. From a state of sheer despair, she is now relaxed, sleeping well, and, for the most part, seizure-free.

At our most recent meeting, some eight years after our paths first crossed, we talk a little more about the impact of her epilepsy on her life. She remains on medication, and is approaching the menopause. Her hormonal fluctuations have almost ceased, and so have her seizures. She has not had any choking attacks for three months; from having many a night, almost every night, to nothing at all. She recalls both the dread of sleep and the completely overriding sensation of being sleep-deprived, of lacking the energy to do anything outside of work.

'My life has changed a lot now,' she tells me.

> Before, I couldn't enjoy life. I tried to with my friends but I just couldn't because of the condition. It would have been nice to be able to go out with friends on a more regular basis. But I couldn't be free and do what my friends were doing. This was forced upon me, and I had no other choice but to limit my life. Now I feel that I have been given part of my life back. I feel like I can mourn the loss of my past life. I am starting a new life and doing the things I want to do.

The sense of rebirth in her is palpable, and I think it is as much a result of restored regular sleep as the resolution of her seizures. But I cannot help but feel sad that she has gone for more than thirty years undiagnosed and untreated. 'I can't be resentful or bitter about what happened to me as a youngster. But I do honestly think it would have helped me so much if I had been diagnosed so much earlier in my life. This is not self-pity, but I've missed out on all the years of my life. I have missed out on so many things.'

In Chapter 5, on Robert and his sleep-talking, I have cautioned about taking things at face value. Janice's story illustrates the flipside of the coin. I suspect that if it hadn't been for her traumatic upbringing, and the behavioural issues, she would have been taken more seriously. Rather than assuming this was part and parcel of her psychological distress, treated with sedatives and antipsychotic drugs, she would have seen a neurologist sooner. The impact of her epilepsy on her entire adult life may have been lessened.

And, for me, the lesson from Janice has been to believe what you are being told, or at least not to dismiss matters. It is difficult in clinic sometimes to unravel the biological from the psychological, and I am sure that I am as guilty as others of jumping to conclusions in the face of someone who is obviously mentally unwell or psychologically distressed. But Janice has taught me to start from the default position of belief rather than disbelief, and to look for clear evidence to support the physical or psychological.

In Janice's case, however, an important factor is the rarity of her condition, and the fact that few people are familiar with it. And therein lies her motivation, her intense desire to tell her story: to prevent others with similar conditions from suffering for decades, like she did.

9

FLOATING EYEBALLS

'I'll never forget this one because this is when I realised there's a problem,' Evelyn says. Now twenty-four, she has finished her degree and is living back at home with her mother. She speaks with a south London accent, and is dressed fashionably. A brightly coloured African print bandana in her hair is a nod to her Ugandan background. She seems happy, confident and relaxed, at least until she begins to talk about her experiences.

'One time at university, I saw my housemate in my face, but the thing about it was that she wasn't even there. She'd gone home for the weekend. So when I saw her the next day, I was like, "What was you doing in my room yesterday?", and she was like, "I've just got back. How was I in your room?" So do you know how crazy that is?'

Evelyn sits and describes this experience with a sense of confusion. For the past five years or so she has been plagued by a range of bizarre hallucinations at night. Since starting her university course, these terrifying phenomena have disrupted her life. 'You're seeing actual things happening, but it's just not real. It's a figment of your imagination. It's your imagination taking you to the wildest places and showing it to you in your face. I've seen demonic figures in my room, and when I see those things I feel like I'm in hell.'

The hallucinations that Evelyn describes do not happen during the day; they are very specific to sleep. 'Normally for me it happens when I'm waking up from sleep. So I could fall asleep for example at midnight, and then an hour later I suddenly wake up.' As she wakes she is greeted by these awful visions. 'One time I saw millions and millions and millions of eyes just looking at me, and they were right there in my face.' She describes eyeballs large and small, like a galaxy of stars floating in the room around her, peering at her. 'These visions are detailed – very, very detailed.'

For Evelyn, one of the scariest events was seeing a very real image of a recently deceased family member in the room with her, 'just looking at me. There was no expression. That one completely put me off going to sleep for a while.'

As if these hallucinations are not bad enough, it gets worse. During these visions, Evelyn feels paralysed. Evelyn shudders as she tells me: 'I'll be lying there, staring up. I can't breathe, I can't move. I try to move, I try to speak, but nothing's happening. The most that will happen when I'm trying to speak is murmurs, so it will sound like I'm mumbling in my sleep. And then I start seeing these really graphic images – crazy, scary images in my face. And then it'll go on and on and on.'

She is clearly traumatised by her experiences. 'The thing is you can't blink, so it's not as if I can shut my eyes and hide from it. It's something I have to look at. And then eventually these things kind of disappear.'

Despite these events happening every night at their worst, she has not got used to them. She has, however, learned how to deal with them. 'I've learned over the months that if I calm myself down – even though no matter how many times it happens to me, it will always be scary to me – the experience will become shorter than previously.' I ask her if she sometimes tries to fight it. 'I try to move my arms first because my feet are just stuck, so I try to move my fingers first but nothing's happening.

And then I try to scream and shout but that doesn't happen. So that's typically when murmurs and mumbling happens.'

It is not difficult to understand why the combination of being paralysed and having hallucinations of people in the room is so terrifying. Evelyn explains:

> It feels like a force is on top of you, stopping you from fighting what you're seeing and feeling. You feel like you're going to die. In that moment when you can't breathe, you can't move, you're seeing things that you don't want to see, dark visions of demons and stuff. You're pinned down, you feel like something's attacking you and is stopping your ability to do all the things that keep you alive.

And, for Evelyn, there is no respite during the day either. She has become so terrified of sleeping at night that she fights sleep constantly, resulting in her being very sleep-deprived. As a result, she has taken to napping during the day, but these phenomena have followed her. She now experiences them even during daytime naps. 'Sometimes it would happen twice in a day when I'd be napping. So when it started happening during my naps, I became even more afraid to go to sleep.'

* * *

It is joked among some of my colleagues that none of my lectures are complete without showing a particular painting by Henry Fuseli, a Swiss artist who lived in the late eighteenth and early nineteenth century. This artwork, entitled *The Nightmare*, features a dramatically reclining young woman, swathed in a long white nightgown, her head dangling over the end of the bed with her arms extended and hanging down. Her eyes are closed and she looks to be deep asleep. What should be a tranquil scene of a beautiful woman sleeping is made deeply disturbing, however, by a grotesque, ape-like demon crouching on her chest in the darkness. Behind the demon, by the woman's legs, floats a horse's head in mid-air, its eyes bulging monstrously in the dimly lit room.

When this painting was first exhibited in 1782, it caused horror and enthrallment in equal measure, and appears to mirror commonly held views of nightmares and folklore that they represent visitations by incubi – male demons who aim to have sex with women in the middle of the night. The concept of the incubus, and its female counterpart, the succubus, were not new, even in 1782. There are references to these demons in Mesopotamian texts from 2400 BC, and these incubi feature in the folklore of peoples around the world, from Germany and Sweden to the Amazon basin, and within many African tribes. Similar descriptions seem to derive from common human experience regardless of race, culture or beliefs.

For me, however, this painting represents many elements of what Evelyn and many other patients describe. The sense of being pinned down, having a weight on their chest, the inability to move, accompanied by deeply disturbing visions of people or ghosts in the room. The horror of this experience is well conveyed in *The Nightmare*, but the universality of this phenomenon suggests that there is a basis in our biology, in our brains, and we now have an inkling of what causes it.

Evelyn has sleep paralysis and hypnagogic hallucinations. As I mentioned in Chapter 6, our understanding of what sleep paralysis is has moved on a little since Mesopotamian days. With the discovery of REM sleep, that stage of sleep during which we dream story-like dreams, accompanied by paralysis of almost all our muscles, we have recognised that sleep paralysis represents the blurring of lines between wake and REM sleep. Imagine the clutch of a car, disengaging as you smoothly switch between first and second gear. When the clutch slips, and the gears grind, switching between gears is no longer seamless. Similarly, sleep paralysis represents the failure to disengage wake from REM sleep, with features of REM entering into full awareness. The paralysis is switched on, and dreaming starts while you are fully awake. In many respects, it is the polar opposite of the REM sleep behaviour disorder that John in Chapter 3 describes, where paralysis fails in REM sleep. It is easy to understand why visions, or sometimes sensory or auditory hallucinations, may encroach wakefulness if the mental processes of dreaming start before you are fully asleep or continue after you wake up. When sleep paralysis occurs during a sleep study, the brainwaves usually show relaxed wakefulness, but the muscle activity is absent, as would be expected in REM sleep.

Very rarely, sleep paralysis with hallucinations has been shown to be caused by epilepsy, but, unlike Evelyn, in those few cases the visual hallucinations are precisely the same each time they occur. While seizures are usually associated with movement, there are little-known areas in the brain called negative motor areas that, when stimulated, induce weakness or paralysis.

Occasionally, I also see patients with very poor vision who have nocturnal hallucinations. In the darkness, the absence of visual signals causes the brain to create visual images. These hallucinations are termed Charles Bonnet syndrome. Patients will experience sometimes simple visions of lights or geometrical patterns, but often highly complex and detailed visions of people, objects, faces or animals, sometimes in miniature – termed 'Lilliputian' in reference to the Lilliput of Swift's

Gulliver's Travels. These Charles Bonnet hallucinations can last for seconds or several hours, and occur in full wakefulness. In contrast to hypnagogic hallucinations, they seldom cause people fear or anxiety, and are very quickly recognised not to be real. But Evelyn is normal-sighted, and experiences her hallucinations when she falls asleep with the light on. Her symptoms are immediately recognisable as the classical manifestations of REM sleep overspill into wake.

Both sleep paralysis and hypnagogic hallucinations are cardinal features of narcolepsy, the neurological disorder that affects Adrian and Phil. In narcolepsy, the switch that controls wake and sleep, and REM and non-REM sleep, is damaged, and patients with this condition often go straight into REM sleep from wake, rather than after sixty to seventy-five minutes of sleep, as in normal sleepers. In these cases, understanding why you may experience these events is plain to see. But why do people without narcolepsy experience this horrible phenomenon?

It is certainly not rare. Many people experience it at some point in their lives. Research in this field is limited, but it appears that certain factors predispose to sleep paralysis and the associated hallucinations. Age, gender and race do not seem to be particularly relevant, but from looking at family history and twin studies, there does appear to be a genetic element. More importantly, however, sleep disruption is a common association. Shift work, night cramps, sleep apnoea and sleep quality in general all appear to increase the likelihood, as do certain psychiatric conditions like post-traumatic stress disorder (PTSD) and anxiety. What all these factors have in common is that they make it more likely for you to enter into REM sleep very quickly, or they result in unstable sleep, and perhaps predispose you to waking directly from REM.

For Evelyn, a sleep study does not confirm narcolepsy, nor does she have the marked sleepiness that is the single most important feature in the diagnosis of narcolepsy. But she does have other factors that may contribute.

When her experiences first started, she was working shifts alongside her university studies. 'I felt like that's when it first happened,' she tells me. 'And then there was a period of time when I didn't have a job, so I was able to just live freely and do whatever I was doing. In that period of time nothing happened.' But it was after university, when she began working at a London tourist attraction, that her sleep paralysis and hypnagogic hallucinations deteriorated. 'When I started there, I only had particular shifts because I was still a student. But especially after finishing [my degree], when you open up your availability and you can say you can work at any time on any day, that's when things became a bit more hectic and a bit more unbalanced. That's when the sleeping issues became a lot worse.'

To compound her problems, the fear of these events made her even more sleep-deprived.

> I began to hate sleep. And I'm a big fan of sleeping. Anybody that knows me knows that I enjoy sleeping. It was very hard because I just didn't want to go to sleep any more. I'd spend a lot of my time awake down here [in the living room] doing anything to avoid going to sleep. I'd try not to lie down because I felt like, if I lay down, I'd end up falling asleep. I'd literally try to make myself as tired as possible. I'd watch series on Netflix. I'd watch films until I knew that I was outright tired. I was happy to go to work and wear myself out knowing that when I came home, I'd be tired.

Rather than helping her get better sleep, though, this was probably contributing to the problem.

* * *

Dreams come in all forms. There are obviously recurring themes, but most people recall dreams of different types. So why should the hallucinations that accompany sleep paralysis, that we take to be the mental processes of dreaming while awake, be so similar and universal? The hallucinations are typically of humans or human-like figures in the room, an intruder standing over the bed, and incubus or succubus coming to seduce in the night, being pinned down by another person. One explanation may be that we integrate external sensations into our dreams. A bang of a door in the house may enter our dreams as an explosion; the dog nuzzling your hand becomes a tiger that you are stroking. So perhaps that sensation of paralysis, of having difficulty breathing due to weakness of some of your respiratory muscles, is also integrated into your dream. You feel as if you are being held down, or that something is sitting on your chest, and this sensation influences your dreaming processes.

Ultimately, any theory remains speculative, but the famous Californian neuroscientist V. S. Ramachandran offers an intriguing hypothesis. During sleep paralysis, people sometimes describe out-of-body experiences, floating above the bed observing themselves sleeping, or sensations of movement or distortion of the body. Ramachandran argues that many of these symptoms relate to miscommunication between parts of the brain that represent where our bodies are in space. An area of the cerebral cortex called the superior parietal lobule contains a representation of our own body. In normal life, motor areas of the brain send out signals to move our body, and these movement commands are monitored by the superior parietal lobule. But during sleep paralysis, there is no movement of the limbs, and no feedback about the changing position or movement of the body. This confusion gives rise to a failure to know where your body is in space.

It appears that this representation of your body is hardwired, however, and does not develop as we mature. The concept of a phantom limb, imagining that your body is completely intact despite the amputation of a limb, with resulting sensation and pain from that no longer existing appendage, suggests that this three-dimensional map of our bodies does not change in life. Amazingly, phantom limbs have been described in people with missing arms from birth, which suggests that this map is so hardwired that it is written into our very genes before we are even born. Ramachandran proposes that the classic intruder hallucination is a projection of this 'homunculus' – Latin for 'little man' and describing this neurological representation of the human body – into our visual world, due to the disturbance of connections between the superior parietal lobule and motor and visual parts of the brain.

He takes the hypothesis one astonishing step further, though, to seek to explain the incubus/succubus hallucination. He proposes that this representation of our own bodies projects to emotional circuitry in the brain and onto visual areas. He and his co-author Baland Jalal write that this network might 'dictate aesthetic visual preference for one's own body "type". This would offer one explanation for sexual attraction/visual preference for certain body morphologies: for example, why humans (generally) are attracted to humans and not dogs, and why pigs prefer pigs as mates to humans etc.'

In support of this, they give the example of people who have an intense desire to have a limb amputated (xenomelia) and who are sexually attracted to amputees. Their powerful feeling that one of their limbs is not part of them is thought to reflect an abnormality in how their homunculus is represented in the superior parietal lobule, and hence they are attracted to people with similar bodies. Jalal and Ramachandran argue that this 'innate and primal sexual affinity for one's own "hard-wired" body image' may explain why hypnagogic hallucinations will often be sexual in nature.

There are also links between sleep paralysis and lucid dreaming, the process of dreaming while being aware of dreaming. During these dreams, people maintain some insight, can exert a degree of control over their dreams, and can access waking memories. Both can be considered states during which greater or lesser degrees of REM sleep are intermixed with wakefulness. In fact, people with sleep paralysis often describe lucid dreaming as well, and evidence from studies confirms this association. Evelyn reports at least one episode of what could be considered lucid dreaming – a dream within a dream.

'This is going to sound crazy,' she begins, 'but I fell asleep and I dreamt about the exact situation I was in, which was me sleeping on the sofa. So my dream was a reality. It was just like normal life was going on, and then in that dream I had sleep paralysis and I woke up from it in the dream.'

In particular, lucid dreaming is strongly associated with out-of-body experiences in sleep paralysis, rather than intruder or incubus hallucinations, leading some researchers to propose that these former phenomena are a feature of positive emotions associated with dreaming imagery, as opposed to the scary negative emotions of hallucinations involving strangers in the room or sexual assault.

* * *

When I visit Evelyn at the family home, the walls are covered in beautiful African batiks, with Swahili banners, family portraits and religious art that represents the family's Christian beliefs. I sit on the sofa, chatting to Evelyn and her mother, and ask what they both made of her sleep paralysis when it first began to happen. 'Me and my mum thought it could be some sort of spirit in the room or it could be somebody trying to put a curse on me,' Evelyn tells me. 'You just don't know what to think. So originally we prayed on it, we put holy water on my bed.'

Evelyn's mother confirms this: 'I said straightaway: "We need to pray and see whether your problems will fade away." And I remember we had a priest who was visiting from Uganda. He prayed for her for nearly half an hour, saying she had some issue with her.'

People around her were telling her something similar.

> We've had this conversation so many times. There are films on TV, Nollywood ['Nigerian Hollywood'] films, that focus around black magic – juju as people like to call it. And especially with something like this, there's often a conception where if you tell people you're seeing things, especially coming from a very cultural background – whether you be African or Caribbean or Asian – a lot of people will link it directly to black magic or juju. And they'll say somebody's put a curse on you. So when you tell people things like this, often people will say directly, 'Oh, it's a spiritual thing.'

Given the nature of these experiences, it is easy to see why. Visitations of people, sometimes dead relatives, or demons at night, out-of-body experiences, being pinned down – these experiences all have a supernatural flavour, and even for people who are not religious, it is thought that these hallucinations may explain reports of alien abductions at night. Evelyn continues: 'I've had a lot of people tell me, "Are you sure it's not juju? Are you sure it's not somebody putting a curse on you?"'

Evelyn tells a slightly creepy story of a bus journey. She had been chatting to a friend on the phone, telling her about her sleep issues. A woman sitting in the row in front had obviously been eavesdropping.

> Then this woman turned around and gave me a note that said: 'This is really dangerous. If you practise this [referring to the hallucinations as an active spiritual practice] then you shouldn't, because sometimes people get caught up in practising it and they become trapped. This thing remains with them for ever.' And I remember I had to cut the phone call off. I said [to my friend]: 'I've got to go. I need to talk to this woman.' And then she [the woman] was telling me how you have to be careful. She told me that apparently her brother used to practise it and it became a problem. It was almost like he was leaving his body and it was becoming an out-of-body experience. According to her, people who practise it, their souls never return to their body and it becomes something really dangerous. 'You have to pray!'

But the power of prayer, holy water and the attentions of the family priest did little to treat her sleep problems. Evelyn recalls:

> Then when we realised nothing was happening, I also had to open up to the perspective that it might not just be spiritual; it could be a genuine sleeping problem. For months and months, I tried to figure it out. It wasn't until

> I came across a video on Facebook that somebody had shared talking about sleep paralysis, which was when I figured out this is exactly what I'm going through. And that's how I knew to go to my doctor and say, 'I think it's sleep paralysis.' Otherwise I wouldn't have known what to say.

Evelyn is philosophical about other people's supernatural explanations for her symptoms.

> You can't blame people for thinking that it's something spiritual or somebody putting a curse on you, because you are genuinely seeing these things. That was my initial assumption. But when I did my research I became a little less ignorant to the situation. I understood that it's not something spiritual. It could just be something that is an actual genuine issue in life. So this is bringing awareness to something that happens to a lot of other people as well.

Simply understanding the nature of her symptoms has had significant benefits. Now that Evelyn knows what is happening to her, the terror that accompanies these events has lessened, although not entirely disappeared. Ruling out a diagnosis of narcolepsy has been a comfort too. But knowing that the key to improving her condition is the improvement of her sleep patterns and sleep quality has opened an avenue of treatment. There are drugs to treat sleep paralysis and hypnagogic hallucinations. Antidepressants suppress REM sleep and delay its onset, and are often a very effective way of dealing with the issue. As a general rule, however, I shy away from starting people on medication, particularly young women who may eventually want to get pregnant. And if there are non-drug-based alternatives, these should be explored first.

Evelyn still does shift work, and this still disrupts her sleep, so it is not surprising that her events are ongoing. But her quality of sleep has improved, and she has become more aware of the need to keep her sleep patterns regular. As a result, whereas previously her paralysis and hallucinations were happening nightly, they are now only occurring about once a month, and she is due to commence a psychological treatment called cognitive behavioural therapy for sleep, to further improve her sleep quality.

* * *

It is worth remembering that sleep paralysis is incredibly common. I have experienced it on one occasion, after a terrible flight back from Australia, sleep-deprived and jet-lagged. Once was enough for me. Even knowing exactly what it was, I remember it as a deeply unpleasant experience. Like so many areas of sleep medicine – as with Jackie and Alex's sleepwalking, and Phil and Adrian's cataplexy – Evelyn's symptoms (and my own sleep paralysis) relate to a failure of normal regulation of sleep. Once again, it is when different parts of the brain are clashing, when the whole brain is not in the same stage of sleep or wake, that symptoms arise. The normal mechanisms of sleep spill over into wakefulness and therefore into consciousness. Understanding this means understanding our brains and how they function, and provides opportunities to treat or even cure these sleep disorders.

10

JEKYLL AND HYDE

The first night Tom* and Sarah* spent together, even before their relationship was properly established, it was obvious that Tom's sleep was not entirely normal. Sarah recalls him suddenly getting up in the middle of the night, popping his trousers on, and saying he was leaving. 'He had no top on,' Sarah recalls, and then describes him going back to sleep. 'He didn't remember anything about it in the morning. So, from very early on, I knew that there was something very strange going on.'

Sarah and Tom had met at a party, and had instantly found a connection. Tom has probable mild Asperger's and can be a little shy, but clearly bonded with Sarah, who is considerate and calm. They look well matched. Tom is slim, now in his forties, tanned and fit-looking. Sarah is of a similar age, well-dressed, with long dark hair. Both had young children from previous relationships. Their own relationship soon blossomed. There were a few times when Tom had initiated sex in the middle of the night, but Sarah had assumed he was just 'being a bit frisky'. It was some three months into the relationship that things took a darker turn.

'We'd been to a party and we'd both had a drink, so I was really, really deeply asleep. I was woken to him trying to put himself inside me through my underwear,' Sarah winces in recollection. 'I was really furious at having been woken up, then been upset and him gone straight back to sleep. I was sore and angry and wanted to talk about it the next morning.' But Tom claimed to have no memory of this and became very defensive. Sarah says: 'He just didn't have a clue what I was talking about.' Her first reaction was horror, and she thought about ending the relationship there and then, but eventually was persuaded otherwise.

Tom recalls the incident vividly. 'I felt physically sick,' he says, and hesitates before he continues: 'It was almost like having an out-of-body experience in that moment. I wanted to punish myself for doing what had happened and what Sarah had described to me. I have a very protective nature about people as well and I think that made it worse. I was looking at myself as this abhorrent kind of creature that was not worth anything because of what I put Sarah through.'

Matters settled for a little while, and all calmed down. But a few months later, a similar thing happened. It was more obvious this time to Sarah that Tom did not seem fully himself, and she was more prepared to believe that he had no recollection of his actions. Despite this, it was clearly incredibly distressing for Sarah. Over the next few months, Tom's nocturnal sexual behaviour would surface every so often.

Tom looks down at the ground and says: 'The only way I would initially know about it was that Sarah would be really dismissive of me, really upset and really angry. It was tangible and it would take me quite a time to actually get out of her what had actually happened. At that point, then she would sit down with me and discuss in great detail what had actually happened.'

With the passage of months, Sarah became more convinced that Tom was actually still asleep. She tells me:

> Over time I realised, because he doesn't do the things that he does when he's asleep when he's awake. It's very distinct from his behaviours when he's conscious. Without the awareness of you having underwear on – and just thrusting, aimlessly thrusting and not really having any goal in mind. Just in an animal way. Never violent, never anything aggressive, just clumsy and awkward and annoying – very annoying, but certainly no malice involved.

She stresses the contrast between his daytime and night-time actions. 'I don't know how to put this. He's not a highly sexed person, so it's very out of character.' There were further clues that he might actually still be asleep. 'I've said to him, "Are you awake?", and he's actually said "yes" before. And I will say, "Are you sure?", and then he doesn't reply, because he isn't awake and it's just an automatic response. And then in the morning I've said: "Do you remember that you tried to do such and such?" and he hasn't got a clue.'

Sarah's realisation that Tom was trying to engage in sexual activity in his sleep has led them both to the Sleep Disorders Centre, Guy's Hospital. When I first meet them, Tom has already been admitted for a night. He has been wired up with electrodes monitoring his brainwaves, his breathing, his heart rate and leg movements. He has spent the night sleeping in a hospital room, under the watchful eye of our sleep technicians, observing via an infrared camera mounted on the wall opposite his bed. When we review the results, over the single night we see several sudden awakenings from the deepest stages of sleep. This is the signature of non-REM parasomnias, the range of behaviours encompassing sleepwalking and night terrors, conditions experienced by Jackie and Alex in Chapter 2.

Perhaps more unusually, it is very clear that Tom has a remarkable propensity to exhibit both waking and sleeping activity at the same time. There are periods in the night, usually before these awakenings, where the slow delta oscillations of the brainwaves, the fingerprint of deep sleep, are superimposed with much more rapid alpha waves, typically seen in wakefulness. Tom's brain is obviously asleep and awake at the same time, sometimes for up to a minute at a time. This is something we see rarely in adults in the sleep lab, and firmly confirms that Tom has a predisposition to non-REM parasomnias. Having read a lot on the internet, Sarah is not that surprised to hear the diagnosis I give Tom – sexsomnia.

* * *

As we've already discussed, non-REM parasomnias come in many guises. These conditions arise from an underlying predisposition not to wake fully from deep sleep. It seems that when deep sleep is disrupted, different parts of the brain wake up differentially. For people prone to this, the parts of the brain controlling movement and emotion are more likely to wake fully, while the regions of the brain influencing rational thinking and memory remain asleep. During this state, sufferers can do any number of things.

Sleepwalking is the best known type of this phenomenon, but other familiar forms include sleep-talking and night terrors. I have seen people eat, cook, rewire household appliances, urinate and, as has been described in an earlier chapter, drive cars or ride motorbikes in their sleep. Occasionally this manifests as 'confusional arousals' – people who, when woken, remain markedly disorientated or altered for some time. In some patients, they exhibit multiple or unusual forms. I remember one unfortunate young woman who lost more than one boyfriend due to her standing up in the middle of the night and urinating on the bed, usually with her partner still in it. Rarely, however, people have been known to have sex in their sleep. Their non-REM parasomnia expresses itself in the form of sexual activity, known as sexsomnia.

Sexsomnia, in the broadest terms, describes any behaviours of a sexual nature arising in sleep. The sorts of activities include fondling, speech of a sexual content or groaning, masturbation, pelvic thrusting or attempted or full intercourse. Sexual activity has rarely been described in the dream enactment that typifies REM sleep behaviour disorder (Chapter 3), or even due to epileptic seizures. Epilepsy resulting from the sensory cortex has been reported to cause genital sensations resulting in orgasm, and seizures arising in the frontal lobes have been shown to cause pelvic thrusting and grabbing at the groin, all arising from sleep. For the most part, however, it is viewed as a non-REM parasomnia.

Like other non-REM parasomnias, amnesia in the morning is typical. While many patients do have evidence of other types of non-REM parasomnia, not all do. There is a huge disparity between the genders – 60–80 per cent of patients are men – and it usually starts in the twenties and thirties. Its rarity may be genuine, but it may also be that people with this condition do not come forward often. At our centre, we see about 3,500 new patients with sleep issues every year, but there have only been about forty cases of sexsomnia. This percentage is borne out by published data from other sleep centres, too. Nevertheless, after a national news story on this condition, I was inundated by emails or tweets from people, men and women, saying that they had it too. Indeed, one study found that almost one in ten people in a sleep clinic setting reported sexual activity in sleep.

Tom's sleep study is absolutely characteristic of that seen in sexsomnia. These sudden partial awakenings in deep sleep, sometimes associated with an EEG pattern of fast and slow brainwave rhythms simultaneously, are typical of those cases in the medical literature. It is rare to actually witness sexual activity in the sleep laboratory, but perhaps this is related to the fact that patients having sleep studies are in bed alone, while incidents of sexsomnia are thought to be frequently triggered by the touch of a bed partner, or something else external causing a partial awakening from very deep sleep. For this reason, sexsomnia is often viewed as a confusional arousal, a period of altered behaviour or confusion related to being incompletely awoken from very deep sleep.

For most people, sexsomnia is not necessarily a major problem, if it happens with their long-term partner. (Perhaps this is one of the reasons we see it rarely, and also why we see it more commonly in men than women; it may be that women, like Sarah, are more likely to encourage their partners to seek medical help.) But when it happens with someone who is not a regular partner, or, worse still, a stranger, it can have devastating consequences, life-changing for both the sufferer and the person sharing their bed. And, indeed, this explains why Sarah has so tenaciously pushed Tom to see a doctor. Because, a few years ago, long before she met him, Tom was convicted of rape of his ex-partner. He was tried by a jury and sentenced to seven years' imprisonment.

Sarah knew of Tom's past long before they were a couple. 'When we first met, he was just staying with me, there was no relationship, but he told me immediately what had happened before with his ex-partner.'

Tom tells me the story of the night in question.

He had had a daughter with his ex-partner, and would stay at his ex-partner's home to spend time with her at the weekends. The distance between his home and his daughter was simply too far to visit for a day. One night, he and his ex-partner were watching a film, and had opened a bottle of vodka. Tom continues: 'My ex-partner said that she was going to bed because she had to be up early the next morning. So she went to bed and I carried on watching the film for probably another half an hour. At that point I realised I was tired, so I went up to bed. I got into bed naked and my ex-partner was wearing a thong. And basically I went to sleep. I think it was something like eleven o'clock at night.'

Tom recalls drifting off to sleep very quickly. 'I was woken up about half an hour, forty-five minutes later to my ex-partner screaming at me and shaking me, and basically telling me to get off and "Stop it, you're hurting me."'

I ask him if he remembers doing anything.

> No, no, no, not at all. And she was screaming at me, 'What are you doing? This is not like you, this isn't you,' over and over again, repeatedly. At this point I remember going downstairs and she followed me down and an argument ensued and that argument became a little bit violent. Not on my part – on my ex-partner's part. She was pushing and pulling me around, pulling me by the hair and screaming at me. I had no idea what was going on and I was completely in shock.

Tom tells me he became very defensive, and fled the house. 'I walked to the train station just completely confused and shocked. I found myself sitting at the train station at about one in the morning, waiting for the first train back to London at half-past five.'

Tom heard nothing more from his ex, until two weeks later. He was at a work-related event in another part of the country. He grimaces as he remembers that day. 'While at that event, I think three days into the week-long event, the police turned up. He was actually a very nice local policeman who hadn't got a clue what he was taking me away for. He just had instructions to find me and arrest me [for rape] and take me to the local police station.'

After an initial trial that collapsed, Tom was eventually convicted of rape at retrial, and served three and a half years in prison, with a further three and a half years on licence.

It was Sarah who first linked Tom's night-time behaviour to his conviction. Over the years she has become an amateur investigator. On reading through the transcripts, she is struck by a very familiar story. Sarah says:

> It was very clear to me, reading what she [Tom's ex-partner] said had happened, that he'd sort of rolled onto her and she'd tried to wake him up and she was saying he wasn't responding. It just was absolutely clear to me that it had been one of his episodes and that if they'd talked about it . . . Like she said, it's not him, and she was trying to communicate to him what he'd done but didn't use those words. If they'd have talked properly, it would have soon become apparent what had happened, but obviously there wasn't much awareness about it [sexsomnia] in those days.

She also tells me that Tom's ex-partner reported him trying to penetrate her through her knickers. 'She had knickers on. It gave her a cotton burn. He just did what he has done to me.'

* * *

There is a reason we are fascinated by stories like *Dr Jekyll and Mr Hyde* and *The Incredible Hulk*. These stories encapsulate the duality of good and evil that resides within us, and the possibility that we all may have a hidden dark nature, an inner 'Mr Hyde', shocks and intrigues us in equal measure.

For Tom, however, this realisation has had a huge impact. For over a decade, he had always assumed that his ex-partner had made the whole incident up, some twisted attempt at revenge for perceived wrongs. I can't help but feel that the anger he felt towards his ex-partner somehow helped him cope with what has happened to him. But the diagnosis of sexsomnia, and the realisation that in all likelihood he did do something on that night, has shaken him to his core. He has had to acknowledge that there is something within him, something dark that arises at night, that he has no control over. Sarah's analysis is very poignant:

> For years he obviously had assumed that it was all lies, that he'd been accused of something that he just hadn't done. He was very confused, very traumatised by the whole thing, and of people thinking that of him. And then to come to terms with the fact that he had actually done something that he had no recollection of whatsoever, and dealing with that sort of guilt after the hatred that there was before. Knowing that about himself and sort of reappraising himself on the basis of that knowledge.

It is important to stress that Tom was found guilty in a court of law by a jury, and in the eyes of the law remains guilty until such time that he chooses to appeal his conviction. I am not a judge or lawyer, and, not being privy to all the testimonies, I am not in a position to say that his sexsomnia was responsible for the event in question. Based upon my clinical judgement, what Sarah describes and what the sleep study shows strongly points towards a diagnosis of non-REM parasomnia, with sexsomnia as one manifestation in Tom's case. In fact, it is clear that Tom has other parasomnias too. Sarah's description of their first night together, when he stood up and went to leave the house partially clothed, is typical, as is a very recent event that Sarah describes.

> The last one [event] he had was when his daughter came to visit, so he was quite stressed about having to look after her. He got up in the night and thought he was on call and was going into the hospital because they needed him. He pulled himself up and got up onto the side of the bed and sort of sat up and told me he was going into the hospital. And I said: 'No, you're not. Go back to sleep.' And he didn't remember anything about that the next day. But it was the stress of having her there.

(Tom is not a doctor, nor does he work in a hospital, by the way.)

Determining that someone has a sexsomnia is in itself not enough, however. Proving that someone was in the midst of a sexsomnia event while undertaking an alleged crime is problematic, and remains a hugely controversial and challenging area of forensic sleep medicine. Tom is not the first person with sexsomnia to run into problems with the law. One of the earliest examples was a case from 1897, when a man was prosecuted for exposing himself while sleepwalking. Over the years, though, there have been several cases where sexsomnia has been used as a legal defence. Critics point to this being a convenient defence for sex offenders, and the major issue is determining the association between a sexsomnia and the event being prosecuted for. Without having EEG electrodes attached at the time of the alleged offence, it is impossible to be definitive. Nevertheless, there have been a number of cases where an acquittal has resulted from a diagnosis of sexsomnia. In fact, it is not just sexsomnia that has been used as a legal defence. Sleepwalking, too, has a long medicolegal history, and may have some bearing on criminal acts. Reports of violence in sleep date back hundreds of years. In medieval times, a woodcutter from Silesia – then in Germany, now in Poland – was reported to have woken in the middle of the night, picked up an axe and aimed it at an imaginary intruder. He awoke to find that he had killed his wife.

Charcot, one of the founding fathers of modern neurology, was asked to pass medical judgement on a servant who had injured his landlady and another member of the household by firing a gun shortly after going to sleep, in an apparent case of sleep-shooting in 1893.

Perhaps one of the most famous, or infamous, cases is that of Kenneth Parks. At the time of the incident, Parks was living in Pickering, Ontario, and was a happily married 23-year-old with a baby daughter. In the early hours of 24 May 1987, Parks apparently got out of bed, got dressed (without socks or underwear on), and drove 23 km west along the shore of Lake Ontario, to the house of his parents-in-law in nearby Scarborough. His first proper memory after going to bed that night was being in a police station, saying: 'I just killed someone with my bare hands. I just killed two people.' It materialised that Parks had gone into the house, had taken a tyre iron and some knives from the boot of his car, stabbed and beaten his mother-in-law to death and had strangled his father-in-law to unconsciousness, stabbing him too for good measure. The story took a further twist when it came to light that Parks, in the year preceding this event, had developed a gambling addiction and had raided the family coffers and embezzled money at work to cover his massive losses. He was due to go to court in proceedings taken out by his employer. Despite a reportedly very close relationship between him and his in-laws, his claim that this must have happened in his sleep was for obvious reasons sceptically received, by doctors and the judicial system alike.

But, despite numerous attempts to confuse him, his story was incredibly consistent, and a subsequent recording of his brain activity in sleep was said to be highly abnormal, and consistent with a non-REM parasomnia. His wife told the court that he was a very deep sleeper and was frequently incredibly difficult to wake. He had also been known to sleep-talk and, on one occasion, sleepwalk. It was evident that there was a strong family history of various non-REM parasomnias. He had been profoundly sleep-deprived and anxious as a result of the pending court case, and evaluation by several sleep physicians and psychiatrists failed to provide an alternative explanation. Rather amazingly, he was subsequently acquitted.

As a neurologist, I have certainly seen people exhibit very unusual behaviour, aggression or violence when there has been brain dysfunction. Patients lashing out when their blood sugars are very low, others frankly but briefly psychotic after a seizure. No one would argue that this sort of behaviour reflects an underlying moral weakness or a failing of one's personality; rather, it is simply a disorder of brain function. Behavioural change is common after brain injury, like the famous case of Phineas Gage I described in the introduction.

One person who sticks in my memory is a man I saw with a serious brain injury as a result of assault. Having been a regular churchgoer, married and running a small business, after his assault he turned into a cannabis-smoking menace to society, racking up ninety-seven convictions over a matter of a few years. While it is easy to understand abnormal behaviour in those cases, where there is something clearly wrong with the brain – a lesion, damage or something altering normal brain function – instinctively it is much more difficult to comprehend this as a phenomenon simply arising from sleep. But this is a lesion of the brain, just like Gage's tamping iron passing through his frontal lobes, or the damage inflicted by an aggressor. It is electrical, or functional, rather than structural, but it is a lesion nevertheless. The intrinsic brain structure, the neurones, the pathways and connections, are unchanged, but the way the brain functions as a whole is temporarily disturbed. While bits of the brain work normally, others do not, and in thankfully rare scenarios this permits these bizarre, violent and dangerous actions to occur – the ability to walk, talk, fight, stab, shoot a gun or have sex in the absence of conscious thought or rational thinking.

So, from a medical perspective, extrapolating from other neurological conditions, most doctors would comprehend the possibility that it is possible to act out violently in sleep. But how does the law view this? In most jurisdictions, criminal responsibility relies upon demonstrating two mandatory elements: that you have performed the criminal act, termed *actus reus*; but also that you have a guilty mind, *mens rea*, meaning that you had a conscious desire to perform the act. For most of these cases reaching a court of law, the physical act and who it was performed by are not in doubt. Assessing for *mens rea* is much more problematic, however, and it is ultimately down to the court, guided by medical experts, to determine. To have *mens rea* requires an intent to perform an act, plus an understanding of the nature of the act, and of the consequences of that act.

The legal arguments therefore largely centre on whether someone was in a parasomnia episode at the time of the act, as this is inconsistent with full consciousness.

From a legal perspective, therefore, parasomnia, and thus also sexsomnia, is viewed as an 'automatism', and can be used as a defence. Automatism is a state defined by grossly impaired consciousness, where a person can perform acts devoid of conscious will, where someone is acting like an 'automaton'. The act itself is involuntary, unconscious. But here it gets even more complicated and messy. If you know that drinking alcohol triggers your parasomnia, this may be viewed as being self-induced, and as such is not a valid legal excuse. Furthermore, in countries that practise common law, like the United Kingdom and Canada, automatisms can be classed as insane or non-insane. In this context, the term 'insane' does not imply a psychiatric disorder. The important issue is whether the automatism has arisen from internal or external factors. If the automatism has arisen from an external factor, like a head injury, or a drug prescribed by a doctor, factors that are not foreseen and unlikely to happen again, this is viewed as non-insane. If it arises from internal problems like epilepsy or sleepwalking, the automatism is insane. Of course, sleepwalking can be triggered by external factors, like noise, abnormal stress or drugs, and this is where the legal arguments become grey. But the distinction is important. An insane automatism, implying a disease of the mind, can in theory lead to indefinite detention, as ultimately the underlying cause of the act remains in perpetuity, whereas a non-insane automatism related to an external factor can be prevented. In the UK, however, an insane automatism defence can lead to supervision orders, where you remain in outpatient medical care, or, rarely, even an absolute discharge, depending on the judge.

But the law remains somewhat opaque on many of these issues, and this is reflected in the range of outcomes seen in the courts when these defences are used.

For the sleep expert, there is some relief that the decision as to whether an automatism is insane or non-insane is for the judge and jury to make. But, from a clinical viewpoint, there are some important questions to ask in order to determine if a parasomnia could explain a violent or sexual assault. There are no universally accepted guidelines, but some features clearly point one way or the other. Factors that strongly point away from a parasomnia would be evidence of planning, seeking out a victim or sexual partner, a clear memory of the event, and attempts to cover up an illegal act. Equally, navigating unfamiliar environments, having a clear motive, and the act being in keeping with the person's character support the view that this was not a parasomnia. Being previously diagnosed with a sleep disorder, providing descriptions of the event in keeping with previous events, confused behaviour, difficulty navigating around obstacles like furniture, or shock and horror upon discovery of what has occurred would all be more indicative of a parasomnia explanation.

For Tom, there is no denying that his sleep study is very supportive of a parasomnia, and in fact he has spent a further two nights in the sleep laboratory, both of which show the same brainwave signature seen on the first night. Sarah describes behaviour entirely in keeping with a sexsomnia, and there is evidence from her description of other non-REM parasomnias. His ex-partner's statements that his actions were very out of keeping with his character – 'this is not like you!' – are also supportive.

But there are some issues for Tom, most significantly the lack of a preceding history of sexsomnia or parasomnias. Tom says he has not had many relationships and perhaps the only person who can give a good history of his past sleeping patterns is his ex-partner. The more recent diagnosis of sexsomnia has at least given Tom and his ex a possible explanation for that fateful night. Tom hopes that she may help overturn his conviction, and in their most recent correspondence, she has apparently mentioned a previous event that at the time she put down to certain past traumatic experiences.

Another issue is that of his fleeing his ex-partner's house when confronted. My own view is that his response was understandable when put into context. When I met Tom, I considered him to have post-traumatic stress disorder, triggered by an event in his past. He had served in the military in his youth, and was involved in a training incident that led to a friend's death. Due to confidentiality issues, I cannot reveal more details of this event, but Tom has subsequently been seen in a specialist service, where this diagnosis has been confirmed. If Tom is to be believed, and he regained consciousness when he was being screamed at and pulled around by his hair, then the combination of PTSD and his possible Asperger's could certainly explain the strong desire to flee in the face of confrontation.

The other problematic area is the consumption of alcohol by Tom and his ex-partner that night. I and most of my colleagues would view alcohol as being a strong trigger in some people with parasomnias, but this is not universally accepted. The role of alcohol in the area of forensic sleep medicine is hotly contested and has been the subject of very ill-tempered debate in the medical literature. Some experts would argue that it is impossible to differentiate between someone who has woken up drunk and committed an offence and someone in a parasomnia episode.

* * *

The diagnosis of Tom's sexsomnia, and subsequently his PTSD, has at least opened the door to treatment possibilities. 'I can never actually sleep in the same bed with Sarah again, because there's no way I want to put her through this ever again,' was Tom's view prior to the diagnosis. After our initial meeting, he wanted to avoid medication, and so we discussed strategies to avoid further episodes. Things like the avoidance of alcohol, stress reduction and sleep deprivation. Sometimes simply not going to bed naked can help. Tom says:

> Sleeping with your partner in the same bed if you are both naked can have a very profound effect on whether an event or an episode of sexsomnia actually happens. As well as alcohol, as well as being in a stressful environment, being in a stressful workplace, and actually down to being in a very unfamiliar sleeping environment. If we stayed in a hotel for a break or something like that then that could have an impact on whether an episode of sexsomnia actually happens.

He was commenced on an antidepressant drug for his PTSD for a brief period, a medication that is also used for the treatment of non-REM parasomnias. But these caused side effects, and he soon stopped the medication. He has also undergone cognitive behavioural therapy, and has been disciplined about regulating the lifestyle factors that he knows contribute to his condition.

Tom continues: 'I feel very confident now. It's only through your help and Sarah's help that I'm in a position now to be able to be very confident about . . .' He pauses. 'I mean I can never say that an episode of sexsomnia is never going to happen again because . . .'

Sarah smiles, and confirms that there have been no further episodes of sexsomnia for a couple of years now.

* * *

Ultimately, as his doctor, I cannot make a judgement on Tom's innocence or guilt. This is for the court to decide, in the knowledge of all the evidence, if and when Tom and Sarah decide to appeal his conviction. But his story is certainly plausible. And what is clear is the devastation that sexsomnia may have wreaked, for Tom's ex-partner, Tom and everyone surrounding them. For Tom's ex-partner, the diagnosis has perhaps led to the realisation that Tom is not the monster she thought he was, someone capable of trying to rape her in her sleep.

For Tom, the knowledge has shaken him to his very core. He now has to face the fact that, rather than this being a fiction concocted by his ex-partner, as he had always believed, there is something dark within him that he cannot control or completely cure, and that in all likelihood he did what his ex-partner accused him of.

And for Sarah? What about her? She was obviously well aware of what had happened in Tom's past long before their relationship started, but behind her softly spoken façade, I can sense an inner steeliness, a single-minded focus on achieving resolution for everyone concerned. It has undoubtedly taken its toll on her and her relationship with Tom, though. In an email, she writes: 'Efforts to avoid episodes can lead to a loss of intimacy and eventually ruin the relationship. The "precautions" we put in place and his fear of an episode stopped him from going anywhere near me at all for the last couple of years. When you lose that closeness, things eventually fall apart.'

The reality is that, in any one of us, given enough provocation, our brains are probably capable of generating a parasomnia event. The perfect storm – sleep deprivation, anxiety, a bit of alcohol, maybe a prescription drug – could result in something similar. Fortunately, these sorts of cases are incredibly rare, but these basic behaviours, such as fear, violence or sex, reside in all of us, to be potentially unmasked in sleep. As Socrates was quoted, in the ninth book of Plato's *Republic*: '. . . in all of us, even in good men, there is a lawless wild-beast nature, which peers out in sleep'.

Postscript: In a recent email, Sarah tells me that the diagnosis of sexsomnia has contributed to a degree of rapprochement between Tom and his ex-partner. Referring to his incident while serving in the military, triggering off his PTSD, Sarah writes: 'She thinks herself more a victim of war than a victim of rape.'

11

THE WAKING EFFECTS OF COFFEE

I can understand why Freud and other psychoanalysts came to the conclusion that sleep uncovers our hidden desires or anxieties. From the contents of our dreams to our nocturnal behaviours, a common feature is one of primitive, basic themes – sex, anger, fear, violence – intense emotions and impulses. We have already seen these in Tom and his sexsomnia, in Alex and his night terrors, and in John and Evelyn with their terrifying or violent dreams. But there is a further elementary instinct: that of food. No patient I have ever seen typifies this quite so dramatically as Don. And as his case illustrates, in the world of sleep there are many blurred boundaries, including where the psychological and biological start and end.

* * *

When I first meet Don, he has already been diagnosed with sleep-eating. Tall, well-built, with glasses and thinning blond hair, he has a slightly patrician air about him. His American accent remains unaltered despite many years living in the UK. He is now in his sixties, and recalls his problems with food starting in his early twenties.

Don tells me that his childhood was difficult. He grew up in Vermont, and went to school in a boarding school where his father was a teacher. 'It was in a small town, but it was a hilltop liberal artsy school.' It sounds like a place of privilege – 'Bobby

Kennedy sent his daughters there, but he wouldn't send his sons. It wasn't rigorous enough. No Latin!' Don laughs. He lived in faculty accommodation with his father rather than boarding with the other pupils, and as a result felt a little apart from his classmates.

Despite the gilded environment, his youth was troubled. 'Well, my parents were divorced when I was eight and my mother just took off. She was twenty-eight at that point, and she moved to a new state. She left me and my sister, three years younger than I am, with my father.'

He later describes his mother as 'not a nice person' and 'an asshole'. He is scarred by the memory of him and his sister being sent to stay with his mother for a few weeks. At the end of this holiday, his mother sent him back to his father, but kept his sister to live with her.

His sense of abandonment by his mother was compounded by his father, whom he felt very close to. 'He was an excellent father, we had a great relationship. When I was fifteen, he married my French teacher, whom I liked a lot.' However, at the age of seventeen, Don finished school for the summer and went back to live with his father and his new wife; his father having moved on from his teaching position at this stage. 'He was frantically looking through newspapers to find me a job and at one point he took me outside and said: "If anybody asks, I am going to say you are my brother. If somebody says: 'Is that your son?' I will say 'no'."' Don thinks that his father's new wife was unhappy about people knowing that her new husband had been married before, and did not want the existence of her husband's children to become common knowledge. 'He was rifling through the pages of the *New York Times* and he found me a job off in the woods where no one would see me, as a camp counsellor.'

Don was clearly deeply affected by a sense of rejection from both his parents. He says: 'So when I got sent away by my mother and sent away by my father, I figured that if they really knew me, they didn't like me – the two people who knew me best.'

It was at the summer camp in the woods, however, that Don formed a relationship that would influence the rest of his life. He discovered alcohol. 'Once every couple of weeks, we got to go out as a group. I was underage but nobody asked me for ID. The first time I was in a bar, I had heard in the movies "Scotch on the rocks", and I ordered some. I drank eight or ten of those. I thought, *This is just what I have been looking for.* You know – oblivion.'

After that summer, he started a degree, majoring in government at an Ivy League university, but the psychological damage done by his parents took its toll. He soon began to develop some suicidal thoughts and went to see a counsellor. 'To be frank, I showed up [at university] and I didn't crack open a book. I smoked a lot of pot the whole year and gained about 50lbs. Then I realised that I was not going to pass a single exam, so I took off. I did manage to come back one time but I did pretty much the same thing.'

By this time, his father had divorced his second wife, Don's old French teacher. In what smacks of a midlife crisis, Don's father decided to take Don to hitch-hike coast to coast across the United States, finding odd jobs along the way. 'Casual labour, covering swimming pools, when somebody needed something loaded onto a truck, and we picked apples with the migrant workers in Washington State.' Don remembers this time with fondness. But a familiar pattern emerges. They had spent some time in Boulder, Colorado. Don was working in a Mexican restaurant at the time. 'My father said: "I am going to Africa," got on a bus and he was gone again . . .' He later tells me, 'You could say it was a raw wound, if you are looking at this psychologically,' referring to being abandoned for the third time.

It was at that time that Don's sleep-eating became apparent. He was sharing a house with his co-workers and recalls that he and his colleagues had ready money, their wages supplemented by tips that remained undeclared. He does not remember the first time he realised that he had eaten in his sleep, but his housemates would continually notice food going missing. He denied being responsible at first, but soon figured it out when he would wake up feeling very ill.

'There would occasionally be fragmentary recollection, but usually none. There would be evidence like wrappers on the counter. I would wake up and know that I was bloated. And people were complaining about missing food.'

The volumes of food Don was getting through must have been enormous. 'I didn't feel in good health until about 5 p.m., because of the vast amount of food intake during the night. My body simply couldn't process the sheer amount I had eaten. [I was] completely bloated, and [there was] the need to go to the toilet throughout the day. I put on a huge amount of weight.'

Eventually, he would have to do the food shopping for two people, to replace what he would consume at night. 'I felt like I was shopping for the daytime me, but also another person, the night-time me. I knew what that other guy liked: easily edible food, often dairy or yoghurt.'

* * *

Don suffers from sleep-related eating disorder, only relatively recently described in the medical literature, first in 1991. His story is rather classic – involuntary out-of-control eating, often without any recall, and being difficult to wake in these episodes. There are many shared features with sleepwalking – complex activities without any consciousness or awareness. Sufferers will often eat weird and wonderful combinations, sometimes even inedible substances. During certain periods in his life, Don has lived a monk-like existence, clearing his house of anything vaguely edible. He recalls recently trying this and waking himself up with the bitter taste of instant coffee granules as he spooned them into his mouth – they were the only 'edible' substance to hand. He has eaten entire blocks of lard before. When I visit him in his house, Paco, his pet green parrot, sits in a cage behind Don as we chat. Paco feels frustrated at not being included in the conversation, and squawks to interrupt every so often. But even Paco is not immune to Don's nocturnal habits. In the past few months, Don has come down in the morning to find the residue of a bowl of birdseed, covered in salad dressing.

People with sleep-related eating disorder will often exhibit dangerous activities when it comes to food preparation, like making dishes with raw meat, or burning food. I have one patient who has been regularly visited by the fire brigade in the middle of the night, due to her habit of placing plastic plates under the grill, setting fire to her kitchen. Don experiences this as well. In the kitchen of his house, he shows me the melted plastic handle of a coffee jug from a percolator. 'I remember none of this, I can only put it together,' he says. He thinks he must have intended to make some tea in the night. 'I must have been confused and put that carafe on the stove. And it caught fire, the plastic handle, so when I woke up, all of a sudden I realised that flames were coming from here and I had a situation I needed to deal with.'

Then there was the time he had obviously tried to cook pasta in the middle of the night. 'I put the spaghetti in water. It was a deep frying pan and I put it on the stove to boil. I have no recollection of this,' he qualifies, having surmised this afterwards. 'But then I came to on the couch, the water boiled away and the pot just had permanent burn marks on it. Something woke me up and I went [to the kitchen] and the spaghetti was frying there in a completely dry pan.'

The danger comes not only from how food is prepared, but also from the eating and drinking itself. There have been reports of people burning themselves by drinking hot liquids or eating toxic substances. The morning bloating, lack of hunger during the day and weight gain Don describes are all rather typical too.

Sleep-related eating disorder seems more common in people with eating disorders, and is often associated with psychological issues or psychiatric disorders and withdrawal from substances of abuse. But while it can be triggered, once present it is usually relentless, for most occurring every night, sometimes several times a night, and lasts for years or decades. The lack of recall, the absence of consciousness, is associated with the presence of sleepwalking, or sedating medications taken for other reasons. And when captured in the sleep laboratory, the sleep-related eating disorder arises from non-REM sleep, just like sleepwalking and other non-REM parasomnias. So, like sleepwalking, anything that can disrupt non-REM sleep, like sleep apnoea, or the kicking legs of periodic limb movement disorder associated with restless legs syndrome, can give rise to these behaviours.

So, is sleep-related eating disorder simply a variant of non-REM parasomnias, like sleepwalking, sexsomnia, night terrors and sleep-talking? Perhaps, but certainly the evidence suggests that there may be underlying psychological factors in the mix, especially when you consider that over 15 per cent of patients in hospital with an eating disorder do it.

* * *

Since his time in Colorado, working in a Mexican restaurant, shopping for two people's groceries, Don's life has changed enormously. He has had children, he has been married twice, and he has moved countries. He has had periods of terrible drinking, but also years of complete sobriety. He has been in the depths of despair, but also the dizzying joyful heights of new love. He has now been in the UK for almost two decades. The one constant in his life, however, is the nocturnal eating.

I ask him how often it is happening. 'If I had to put a number on it, I would say it is 364 days a year,' he laughs ruefully. Moreover, the sheer drive to eat in his sleep is intense beyond belief. At certain points he has taken to locking food away, sometimes even attaching a bicycle lock around the fridge. 'At one time, I had a girlfriend who became my first wife. In the middle of the night, my girlfriend came out and found me standing on a chair stark naked, trying desperately to reach my arm down through the crack [of the fridge door] to get something.'

Don seems to favour dairy foods, things like cheese and yoghurt. He can easily get through an entire block of cheese or large tub of yoghurt in a night. 'I think basically it has to be easy to eat, so cheese is and yoghurt is easy to eat.' But his psychoanalysts, who he has seen on and off for the past couple of years, have an alternative theory. As a baby, he says,

> I was allergic to milk. I was allergic to my mother's milk, any milk so far as I know. Something for which she [his mother] never forgave me. I am quite serious; she never

> did. But then, at nine months, I guess I was not getting enough nourishment, and I was hospitalised for dehydration. I had been told that you could pinch my skin and it would stand up.

His psychoanalysts have taken an enormous interest in this part of his life story. 'Well, essentially, I was deprived in infancy. You know, they are very big on separation from the primary carer or the primary food giver. They believe that separation from the primary care giver can actually physically alter the brain.' Perhaps this preference for dairy reflects his deprivation of it as a baby.

As we explore psychological drivers further, I ask him about his own relationship with food, but he tells me that during the day this is entirely normal. I ask if there is anyone else in the family with food issues. 'My eldest daughter had an eating disorder. She grew to be morbidly obese and would hide food. It was compulsive eating. She recently had gastric bypass surgery which has helped her return to a healthy weight.'

Don clearly has a huge amount of insight into his psychological make-up, and is certainly not shy of self-examination. I ask him what he thinks has caused his night-eating. Given his openness about the traumas in his life, I am somewhat surprised when he says: 'I think it is a neurological problem. I mean I tell you this other stuff about my background because I can't be sure. But there was the period of time which you referred to as "the false dawn".'

He is talking about a period of one year when, on a combination of drugs, his sleep-eating stopped, only to recur. 'I don't call it the false dawn, because it was a complete cure, albeit it only lasted a year. What that makes me think is that there is some sort of chemical component to it because it just shut off.'

Later, he reiterates a point made earlier: 'Lending weight to the theory that it is something chemical is that it is pretty much a constant. That's not to say that sometimes it is not more severe than at other times. But if I am drinking, if I am not drinking, if I am depressed or in nirvana because I have fallen in love with my second wife, it just carries on.'

Over the years, Don has seen a number of physicians.

> When I was in the States, I went to a sleep centre in Atlanta. He tried various things. He gave me a downer. I then reported that this made it worse. So he then gave me Ritalin or something related to Ritalin, which cured the problem because I just didn't sleep. Well, anyway, nothing worked. Then he grew frustrated, and he said to me: 'Well, you quit smoking, and when you are ready, you'll quit this,' and I was kind of annoyed.

A few years later, he also went to see a sleep physician in Belgium. 'They had never ever heard anything of this sort,' Don says.

One of his major regrets is not talking about his sleep-eating earlier. It was only after many years of suffering that he spoke to a physician about his sleep problems, many years after this condition was described by the medical community. 'I felt a certain amount of regret that I did not discuss it. The only reason it got listed [in the diagnostic manuals] is when people talked about it.' He feels that his silence has contributed to the lack of knowledge about this condition.

* * *

In recent times, Don has noticed a slight change. In addition to his eating in his sleep with little or no recollection at all, which continues unabated, he will sometimes eat at night when he wakes up in the middle of the night. 'I feel fully awake, but the compulsion to eat is absolutely irresistible.' He describes waking up and having an overwhelming feeling that, if he does not eat, he will never sleep again. 'If my wife is with me, she will say, "Lie back down," but five minutes later I will just have to get up [to eat].'

Don's current night-eating syndrome is different from the sleep-related eating disorder that he has always suffered from. In contrast to sleep-related eating disorder, night-eating syndrome is a compulsive act, done in full consciousness, and is not an activity arising from deep sleep. He is fully aware of what he is doing, but simply cannot help himself.

It may, however, be related to the periodic limb movement disorder that I have identified on his sleep study. Whenever he has spent the night in the sleep laboratory, he has been found to have both severe periodic limb movement disorder, on one study kicking 110 times per hour throughout the night, as well as moderate sleep apnoea. His sleep studies have also shown the typical sudden apparent 'awakenings' from deep sleep, with the confusion and subsequent return back to sleep that is seen in sleepwalking. But in addition, we have witnessed some events in which he clearly wakes up, eats a banana, and then eventually goes back to sleep.

Night-eating is strongly associated with restless legs, often existing side by side with periodic limb movement disorder, and it is not simply a matter of occupying waking hours. It is rarer in people with straightforward insomnia, despite insomniacs spending much more time awake. Some researchers point to certain similarities between night-eating syndrome and RLS. In both there is an itch that needs to be scratched, a certain feeling of compulsion. In RLS, there is this urge to move, that builds and builds until there is no choice but to do so. Likewise, in night-eating syndrome the 'itch' is the need to eat, which escalates until the urge to eat has been sated. In fact, treatment of RLS often improves night-eating syndrome. It has been said that standard treatments in restless legs syndrome, the dopamine receptor agonists, may be responsible for night-eating, in the same way that they sometimes give rise to compulsive behaviours like shopping, gambling or hypersexuality during the day. However, there is also evidence to support the use of dopamine receptor agonists in the *treatment* of night-eating syndrome, and that this compulsive eating at night is as a result of the disturbance in the circadian rhythm generated by the restless legs syndrome itself.

Sleep-related eating disorder, without full awareness and out of non-REM sleep, has itself also been associated with restless legs syndrome. And the overlap between sleep-related eating disorder and night-eating syndrome is significant: about half of all patients with sleep-related eating disorder also have night-eating syndrome, suggesting some shared underlying cause. They are distinct – one without recall, often eating foods that would never be eaten during the day or even inedible substances; the other while fully awake, without the tendency to eat unusual foodstuffs. It may be, however, that they are at different ends of the spectrum of a single disorder. An abnormal circadian rhythm when it comes to food, a dysfunctional neurological reward from eating, that manifests in different ways if it arises in wake or non-REM sleep.

So, do I agree with Don that his problem is largely neurological, due to a chemical abnormality in his brain? Well, yes and no. I would argue that he has a tendency to the dual brain state in keeping with other non-REM parasomnias, where part of his brain is awake and other parts are asleep, enabling him to prepare food, to hunt out meals, to undertake fairly complex tasks without any recollection in the middle of the night. And that this tendency is exacerbated by his periodic limb movement disorder, causing partial awakenings from deep sleep, also spending more time 'awake' in the middle of the night, and disrupting his circadian rhythm.

But there is no denying that there are psychological factors at play – his difficult childhood, depression, and his extremely problematic relationship with alcohol. I am intrigued by his daughter's pathological relationship with food, and when discussing Don's alcoholism, I ask if his parents had problems with alcohol too. 'I remember that she [his mother] did like her before-dinner cocktails. Later on I think that she worked. I found out, many years later, after her death, that other than working, she spent about ten years in her room drunk.'

It may of course be that some of the psychological issues have a basis in biology. Don has a tendency to compulsive behaviour, some sort of abnormal reward mechanism, an excessive 'buzz' from certain types of activity, of which the alcoholism is a feature. At one point, when I try to treat his periodic limb movement disorder with a dopa agonist, he develops a compulsion to flirt with women on the internet. This resolves as soon as the drug is stopped. And this dysfunction of the reward system seems to have been inherited from his mother, and passed on to his daughter. So, perhaps there is a heightened reward that Don gets from food.

I am drawn back to his description of his time at Harvard, when he gained 50lbs over a few short months, this happening way before the sleep-eating surfaced. Perhaps that excessive drive for food emerges at night, when asleep, due to the activation of his limbic system while in that non-REM parasomnia brain state, or when awakened by his periodic limb movement disorder. While during the day, he is so distended, so bloated by the vast volumes consumed at night, that he does not eat. Much to my chagrin, I find myself agreeing with the psychoanalysts: that the night-time may uncover 'hidden desires'.

I cannot remember ever having seen anyone with such severe sleep-eating as Don, in terms both of intensity and difficulty in treatment. And I am not the only sleep physician who has tried to treat him. Standard treatments for non-REM parasomnias have failed, and I have tried to treat his sleep apnoea as well as his periodic limb movement disorder. Managing his medications has proved problematic. Many of the treatments for non-REM parasomnias can worsen his limb movements, and I have avoided many of the treatments for periodic limb movements because of their potential for addiction and dependency, of major concern in his case. The dopamine receptor agonist, the standard treatment for restless legs syndrome, triggered further compulsive behaviours. One treatment often used in sleep-related eating disorder is an anti-epileptic drug called topiramate, which acts to suppress appetite. It made no difference at all. At the moment we are trialling iron infusions, the aim being to dampen down the leg movements and address the sleep-related eating disorder and night-eating syndrome in that way. But I do not hold out much hope.

Don has learned to live with his nocturnal appetites. He controls it to some degree by limiting the food accessible to him at night. He has tried to put more healthy food in the fridge, like fruit, but even this can cause difficulties. 'I was eating too many grapes, and waking up with diarrhoea,' he shrugs. And, undoubtedly, his sleep-eating puts a strain on his relationship with his wife, and how he functions during the day. He is frustrated by how the disorder he deals with every night is perceived. 'It is a serious problem that affects your life greatly. I have a very fine career, but that was in spite of the difficulty that I started each day with. I say that because I have more than once seen TV shows on sleep disorders, and each time they include sleep-eating as the comic relief.'

For me, Don's case exemplifies many of the sleep issues that we see day-to-day in the clinic, perhaps not in its nature but in its origins. Sleep is so dependent on biological and psychological factors – your genes, your anatomy, your brain chemistry – but also the stresses and strains of life – your mood, your anxiety levels, your state of mind. It cannot be seen in isolation, only in the greater context of someone's life. And if that is the case with sleep in general, it is certainly the case with disorders of sleep too.

12

A PECULIAR FAIRY TALE

Every few months, I sit in a clinic with my paediatric colleagues, seeing youngsters with conditions like narcolepsy or severe sleepwalking. The aim is to smooth the transition between the 'touchy-feely' world of the children's hospital next door, with its brightly coloured, mural-covered walls and helter-skelter in the foyer, and the slightly harsher medical environment of adult medicine. Some of my paediatric colleagues wear lurid yellow-and-purple shirts, with a different shade of hair colour every clinic, and their ID lanyards are festooned with badges or keyring cuddly toys – not a look that my adult colleagues (or I) cultivate.

And with adulthood come new challenges as a patient. The life of a teenager is more complicated, socially, educationally and medically. For the patient, there is an increasing onus on managing one's own health, and for the teenager and the parents alike, that shift of responsibility can be difficult.

This transitional clinic aims to facilitate the slow weaning of parental involvement, the appropriate handover of care to a new team, and to begin the discussion of the impact of their condition on driving, their social life, alcohol and drugs, and moving away from home. Sometimes, these patients fulfil the stereotype of teenagers – grumpy, monosyllabic, slumped in the chair, devolving responsibility to their understandably overbearing parents who talk for them.

But Jamie is about as far away from that teenage stereotype as one could imagine.

Now seventeen years old, he is articulate, thoughtful and conscientious. He is clearly academic and very sporty. He and his entire family are rugby-crazed. He plays at a very high level, for his school and county; his father coaches; and his mother and his younger brothers are also involved in the sport. He harbours ambitions to study medicine at Imperial or Oxford. The prospect of a career in orthopaedics and sports medicine represents a confluence of his love of sport and his academic potential. But, despite his obvious talents, Jamie has other problems to contend with.

'It started just after my fourteenth birthday,' he tells me.

> I was on a school skiing trip to Serre Chevalier [a French ski resort near the Italian border]. We had a nightmare journey out there. We were caught in a massive snowstorm. The coach was stuck for six or seven hours on the motorway. No food, no drink; forty or fifty of us. No communication. We got to the hotel at two in the morning. Everyone was shattered. But we were there to ski, so at eight o'clock the next morning we got up and went skiing.

The next few days were incredibly strenuous. Jamie and his friends were skiing, ice skating, sledging and staying up late at night. He recalls: 'One afternoon I started feeling really drowsy, so I took the afternoon off skiing. I thought I was dehydrated, as I hadn't been drinking a lot of water. I was drinking a lot of Coca-Cola, which obviously wasn't doing enough. I felt really tired so I fell asleep.'

At that time, he was not on his own. Several other boys were feeling the effects of overdoing it, and some also chose to rest in their rooms that afternoon. But, despite his rest, Jamie continued to feel sleepy, and while his friends went out that night to go ice skating, he elected to stay in bed. 'And two hours later, once they [his roommates] got back from the ice skating, I woke up with the room full of people. They had played a prank. It was something to do with whipped cream, some silly prank that fourteen-year-old boys do. I woke up and I was really disorientated. I felt quite aggressive towards the boys that had done it. I remember just being in a complete daze.'

But, despite not feeling well, Jamie continued to ski the following day. 'I vividly remember going up the ski lift thinking, *Something's not right*, and falling asleep on the lift.' Jamie continues: 'It was a really tricky few days, not feeling well. I just thought that it was some sort of ailment, just some sort of a bug. So I slept in the coach all the way back to the airport. I was really disorientated in the airport, [and] got lost from the group. I didn't know where I was, I didn't know what was happening.'

On Jamie's return home, he remembers being ravenously hungry. 'I sat down and started eating dinner. And I was eating really, really quickly. I remember my family saying, "Just calm down!" They were asking me about the trip, and they were cross with me because they had paid a lot of money for me to go on this trip and I was just silent. I wasn't saying anything.'

At this point, Jok, Jamie's father, takes up the story. 'We went off to rugby as normal on Sunday. We left him at home. We got back in the afternoon and he was still asleep. Jamie was still not rising. We thought maybe he was still asleep because of dehydration; maybe he had bumped his head and had a bit of concussion. But we knew something wasn't right.' Oriele, his mother, describes what he was like: 'Vacant. Big pupils. Just staring, and nothing was getting through, no communication.' It was at this point that Jamie's parents thought it was time to take him to hospital.

In the emergency department, after initial tests all returned normal, the doctors presumed that Jamie was suffering from a psychiatric condition. Oriele recalls them being interrogated by the doctors. 'They were really grilling him about what drugs he had taken. And eventually he said: "I did take a Nurofen because I had a headache,"' she laughs. 'He managed to say that. But then the doctor there said that in the nine years he had been working there, the only time he had ever seen a child like this was when they had been sexually abused.'

Jok still recalls this moment with horror. 'So of course at this point alarm bells were ringing. We were really concerned.'

Eventually, after several hours, Jamie's parents were allowed to take him home for the night, but he was readmitted the following day. Jamie recalls lying in a room on a bed, watching the planes flying into Heathrow overhead. 'I was put on suicide watch. I remember a rather eccentric man looking after me. I was there for a while, eating really horrible food and watching lots of TV, not really doing anything. I don't personally remember sleeping that much. I just remember my family coming in and me manically eating Jaffa Cakes.'

Over the course of the next few days, Jamie had a lumbar puncture, brain scan and an EEG. He also remembers being separated from the family for a psychiatric assessment. 'That for me was really scary,' he recalls. 'Some of the things I said were really uncharacteristic. I was saying I was suicidal and I had thought about killing myself with an aerosol can. I wasn't thinking about what I was saying.' But it was shortly after this that Jamie and his family were given a diagnosis.

The opinion was that Jamie had developed a condition called

PANDAS, a rare autoimmune neurological disorder triggered by streptococcal sore throat. Antibodies directed against the streptococcus bacterium attack a part of the brain called the basal ganglia, and cause psychiatric symptoms as well as some abnormal movements. Jamie and his family remember the diagnosis as a source of relief. He was treated with intravenous antibiotics, and after a few days he gradually returned to normal. After a couple of weeks, he was back at school.

Jok says: 'Obviously the PANDAS diagnosis wasn't 100 per cent clear. We could tell the doctors weren't convinced. But that's what seemed to fit the bill at the time. But by then he was out and he was better and we thought, *Phew!* We had our Jamie back.'

Life for Jamie and his family quickly returned to normal, and they soon put this episode behind them. Jamie continued to study hard and play hard, now playing rugby for a local club as well as everything else in his already busy life. Normality resumed.

A year and a month after his first episode, however, Jamie went to a party with his friends. Rugby players are not renowned for their abstinence from alcohol, and I ask if he had been drinking. 'I have never drunk excessively, but I had had a bit of alcohol, been at this party, and got home quite late. The next morning it kind of hit me again, so I slept all day in a darkened room.' It was immediately obvious to his parents when they got home that something was not right. He was again very sleepy. Oriele says: 'It was so similar to before. It was that vacant stare again. So we took him straight back into hospital. It is strange walking into an A&E department with a fully grown young man – physically fine, but you're trying to explain he is not communicating, not functioning.'

This admission was no less traumatic than the first. He was

kept in a room with doctors walking in and out. 'One doctor, a child psychiatrist, was so rude,' Jok recalls. 'He just walked in, said: "Acute psychosis!", and walked straight back out again.'

For the next ten days, Jamie remained in hospital. His behaviour was strikingly abnormal. '[He was] manically eating,' says Jok. 'We couldn't leave him with a packet of biscuits. CAMHS [child and adolescent mental health services] came in to see him again and he had a packet of biscuits. He ate the whole packet in one fell swoop, which they thought was extraordinary.' Jamie has some memories of this episode. 'I got quite stroppy when my grandparents came to visit. I didn't like anyone unfamiliar. I couldn't speak. I remember speaking to my mum, using Notes [the app] on her phone. I was swearing a lot. They asked me what I'd had to eat or drink, and I just said, "Fucking water!"'

Oriele confirms this: 'Lots of swearing, which is so unlike him. Things like "Bitch!", which he would never say, or at least I hope not!' His parents tell me the psychiatrists soon decided that this was not a mental health issue – 'We are off, we have done our job, we have checked him out, it is not mental.'

It was not until the following day that Jamie began to show signs of other alarming behaviour. Oriele reports: 'Jamie was different, doing strange things around me. It was totally out of character.' Jok chips in: 'It was inappropriate suggestions and things you wouldn't do really around any woman, but especially around your mum.' At this point, I see Jamie shifting uncomfortably in his chair, and throwing a look across to his girlfriend Megan, who by this time has joined us. I can see the colour rising in his cheeks, and feel sorry for his embarrassment. Oriele continues, in a very matter-of-fact way, to describe Jamie's highly inappropriate actions. Jamie winces at hearing the details, and she apologises to him for causing him discomfort. 'I then realised that I needed to keep my distance and not get too close to him.' Jok says: 'So we became quite protective of him and we made sure that invariably I was there, especially when the nurses were coming in.'

* * *

In the strange world of sleep disorders, there is no condition quite as extraordinary as Kleine–Levin syndrome, or as recognisable. Jamie's episodes bear all the hallmarks of this condition, known as KLS, and on our first meeting, just after his sixteenth birthday, I tell him and his family that I am pretty certain that this is what he has. In fact, his parents already have an inkling as to the diagnosis. One of the psychologists during his second admission had said to them that he was interested in neurology, and had read about KLS. He had raised it as a possible explanation, but when Jamie's family raised it with his medical team, the suggestion was met with short shrift.

'We mentioned it to the doctors, but they dismissed it,' Jok tells me. 'But we began reading about it. There were an awful lot of boxes being ticked as we read about the symptoms. The sleeping, the excessive eating, the sexual behaviour, guarded behaviour, derealisation.'

I ask Jamie more about this. 'It feels like I'm drunk. I'm not in my body, like I am almost watching myself. This is really tricky to describe. Like I was in a tunnel and I could see everything that was happening, but I could have no impact on what was outside me. I couldn't feel my hands, my legs, and my feet. I couldn't speak. I could think but I couldn't speak.'

There are very few conditions in modern medicine that we have almost no inkling as to their cause. KLS is one of these disorders. We do know it is rare – exceedingly so. It is thought to affect between one and five in every million people, although it is probably more common in certain ethnicities, such as

Ashkenazi Jews. We know what it looks like, and in fact it has been described for centuries. The first probable case in the medical literature dates back to 1705, in a paper unimaginatively entitled, 'Relation of an extraordinary sleepy person at Tinbury near Bath'. But it was not until the mid-twentieth century that this syndrome – a syndrome being an association of symptoms rather than a disease – was better described.

The earliest modern descriptions, including those by Willi Kleine and Max Levin, physicians in Germany and Baltimore respectively, were of young men struck down by episodes of profound sleepiness, associated with 'morbid hunger' and marked hypersexuality. Since its original description, our understanding of the spectrum of symptoms that constitutes this syndrome has progressed a little. Whereas previously thought to be a condition almost exclusive to males, it is likely that the hypersexuality exhibited in boys is more immediately recognisable and leads to a quicker diagnosis. Nevertheless, KLS is still thought to affect more males than females. It can start at any age, but usually appears in adolescence. Occasionally, first episodes can start in childhood, and the oldest reported patient had his first attack in his eighties, although this is highly unusual. Jamie's age of onset is very typical, as are his episodes.

Patients usually have bouts of behavioural change lasting days or weeks, in rare cases up to several months, recurring every few months. As with Jamie, the usual picture is one of profound sleepiness, so irresistible that patients will sleep anywhere. I have read of patients crawling under cars and being found asleep when someone has driven away, or, as Jamie describes, on a ski lift. One of my patients has gone to sleep on the concrete floor of an airport terminal, surrounded by hordes of passengers. Sleep may constitute up to twenty-two hours of the day, and the degree of sleepiness can be so severe that sufferers may only get up to eat or go to the toilet. I can recall at least one patient whose parents would leave bottles of fizzy drinks next to the bed. The only evidence of him having risen would be that the bottles of cola would be replaced with bottles of urine. When challenged to get up, the young person will often be incredibly difficult to rouse, and when woken will become very aggressive, with violent outbursts of swearing. These episodes of sleepiness have led to KLS being described as 'Sleeping Beauty Syndrome', a headline much loved by the tabloid newspapers, whenever a story on KLS is featured. Patients and their families hate this title with a passion, however, believing that it belittles the effects of the condition on people's lives, and romanticises KLS.

As evidenced by Jamie's story, sleepiness is not the only symptom of KLS. Kleine and Levin noted that, when awake, the patient with KLS exhibits some very bizarre behaviour. First, there is the 'morbid hunger', the uncontrollable urge to eat. I have seen patients eat food that they would not touch in normal life – vegetarian patients demolishing large amounts of meat, for example – and others who have entirely cleared out the family kitchen of all things edible. Often there is a tendency to junk food – sweets, chocolate, crisps. There is a case in the medical literature of a patient eating so quickly and voraciously that they died of pneumonia, having inhaled rather than swallowed some food. Jamie's consumption of a whole packet of biscuits is tame by comparison.

Then there is that most striking of symptoms, the hypersexuality. It is not just males who describe it; female KLS sufferers too, and it is a source of huge embarrassment for all concerned. Occasionally, it can also be dangerous. One of the first patients with KLS I saw was a young man from the north of England who, during bouts, would become so hypersexual that he would sit outside his house and proposition every passing female. When I told him I presumed his success rate was pretty poor, he replied: 'You don't know the kind of neighbourhood I live in!' But, despite his apparent levity, one of the reasons for his seeking out medical attention was that he feared that his behaviour was so extreme, his sexual drive so intense, that he might undertake criminal acts.

Although this triad of symptoms – sleepiness, excessive eating and hypersexuality – constitutes the core symptoms of KLS, we are increasingly recognising that it is actually the minority of patients who exhibit all of these features. Rather than profound sleepiness, during bouts people often feel confused or in a dreamlike state. They often describe a feeling of being in a bubble, as if the world around them is not real. Jamie recalls a sensation of his brain not being connected to the outside world. Apathy is commonly described, as is mood disturbance or anxiety. Many of my patients feel incredibly disturbed by anything unfamiliar, similar to Jamie's response when his grandparents came to visit him in hospital. But it was not just unfamiliar people that Jamie did not like, it was also unfamiliar things.

Jamie's mother remembers him singing the same song over and over, or listening to the same track again and again on his iPod. Many of my patients will watch childish, predictable television over and over, typically cartoons or Disney films, sometimes several times a day, back to back, and like Jamie will sing the same song incessantly. (One particular young woman drove her family to despair with repeated renditions of 'Let It Go' from Disney's *Frozen*.) Anything new or unexpected causes them distress. And the childlike behaviour extends to speech. Patients' families will often report that not only do patients become less verbal during bouts, but the content of the speech will become childlike, with childish or infantile speech patterns. Jamie's parents comment that his intonation will change. 'Just simple words. "Dad-dee! Biscuits!"' Jok reports.

Having now seen several patients in the midst of their bouts, it is easy to see why these events may be viewed as psychiatric, psychological, or someone simply swinging the lead. It is really one of the most peculiar clinical pictures I have ever encountered. Imagine a teenager lying in bed, not really awake or asleep, alternating between being completely mute and unresponsive, and suddenly getting up to eat five chocolate bars in a row, then getting back into bed. I have been high-fived, had laughable or highly inappropriate conversations, have been sworn at. I sometimes feel as if I am talking to a baby inside a seventeen-year-old's body, with conversations about *SpongeBob SquarePants* playing on the television mounted on the wall in the sleep laboratory room, or about a favourite cuddly toy. I have had more sensible conversations with my daughters when they were five years old.

At its most extreme, these bouts of Kleine–Levin syndrome can even be associated with delusions or hallucinations. It is not unusual for patients with KLS to receive an erroneous diagnosis of a psychotic disorder, such as bipolar disorder or schizophrenia, as was the case with Jamie.

So how do we know that KLS is actually a disorder of brain function rather than a psychological or psychiatric disorder? Could this condition just be a case of a lazy teenager seeking ways to escape the stresses and strains of life as an adolescent, as many of my patients have previously been diagnosed?

There are no tests that can be done – no MRI scans, no blood tests, no lumbar puncture – to prove the diagnosis of KLS. The diagnosis is one of exclusion, i.e. all other possibilities need to be considered first. It would be fair to say that the diagnosis of KLS is made when an 'expert' says it is KLS. So, is KLS another example of the emperor's new clothes? Much like Hans Christian Andersen's Emperor parading through the streets naked, believing himself to be dressed in a magic suit that could not be seen by stupid people, applauded by his subjects, also not wanting to declare themselves stupid? A case of 'an expert calls it KLS, so everyone believes it to be KLS'?

In my own practice, I try hard to exclude all other possibilities. All my patients will undergo a psychiatric evaluation to rule out an alternative diagnosis. There are some similarities between bipolar disorder and KLS, with patients cycling between mania with associated insomnia and depression with prolonged sleep, but I would hope between myself and my psychiatric colleagues we should be able to distinguish the two.

Other conditions that can mimic KLS include very unusual forms of epilepsy, genetic disorders that influence metabolism, and the equally rare Klüver–Bucy syndrome – where damage to the temporal lobe on both sides of the brain also results in excessive eating and hypersexuality, but also amnesia, difficulty recognising visual objects and placidity. Sometimes the explanation is more prosaic, in the form of the ingestion of illicit substances. All my patients have a battery of blood tests and a brain scan.

Even so, it does ultimately come down to a diagnosis of exclusion, and the recognition of a pattern of symptoms. Much of medicine is about pattern recognition – realising that the patient in front of you sweating profusely with a racing heart, clutching his chest, is having a heart attack, much like the ten patients you have previously seen who looked exactly the same. With KLS, this is much harder. Recognising a pattern requires you to have seen several similar patients before – extremely hard in a condition that affects perhaps one in a million people. And so even 'experts' will have seen relatively few patients with this condition. As a junior doctor, I used to look at my senior colleagues, experts in a particular condition like lupus or sarcoid – old favourites of the scriptwriters for the medical TV series *House*, due to their myriad presentations and problematic diagnoses – who appeared to have developed a personal relationship with their chosen disease. They knew the personality of a disease, its foibles or idiosyncrasies. It was almost as if the science of the diagnostic process for them had been replaced by intuition, a feeling that, despite all the tests being negative, this *is* lupus after all. Having seen roughly forty patients with KLS now, I would not claim to have that degree of closeness, that intuitive relationship with this condition, but there are little trigger words, common phrases, familiar descriptions of symptoms by the patient or their family that approach this level of intimacy with KLS for me. I hope that I recognise it when I see it, but it is not always easy. There are several patients in whom, even after a couple of years of knowing them, I am reluctant to give a firm diagnosis of KLS, preferring to keep the door open to an alternative diagnosis that may be easier to treat.

Even so, what I am diagnosing is a syndrome, not a disease. It is purely an association of multiple features, a recognisable pattern of symptoms. Just because there is commonality between more than one patient does not confirm a biological cause of brain dysfunction for this syndrome. I have seen countless patients with numbness, tingling or paralysis in a very recognisable way that have an underlying psychological basis. But, actually, there is some evidence of changes in brain state in people with KLS.

The EEG – analysis of electrical activity – shows abnormalities in the majority of patients during bouts that disappear in between bouts. While MRI and CT scans are normal in people with KLS, scans measuring brain blood flow or brain metabolism of glucose, either using radioactive substances or MR scans that monitor blood perfusion, show alterations in activity most consistently in the thalamus, but also in other brain areas.

These changes are evident even between these bouts of sleepiness and behavioural change, and, indeed, recent evidence suggests that these brain changes are accompanied by subtle difficulties with memory even between bouts. In addition, assessments of cerebrospinal fluid during and in between bouts have shown lower levels of hypocretin, the neurotransmitter implicated in narcolepsy, during periods of sleepiness. These abnormalities could clearly not be explained by someone with psychological issues. These are as close to diagnostic tests as we have, but are insufficiently specific to use in routine clinical practice.

* * *

The relief of a more certain diagnosis was dramatic for Jamie and his family. Oriele recalls that their first thoughts were that, now they knew what it was, and that it was not life-threatening, they as a family could get on with life and deal with it. Jok also remembers feeling optimistic. After all, Jamie had already gone for over a year between his first and second episodes. Their initial positivity soon evaporated, however, as Jamie began to have episodes every few months. Jok says:

> The first four or five [episodes] were very distinct ten-day episodes and there was very much a pattern within those ten days. The first two or three days were very deep sleep, hardly any interaction. He would be almost catatonic. We actually set up a baby monitor because he wouldn't come

out of his bedroom, he wouldn't get up. He would only wake to eat and he would be very demanding, just saying, 'Food!' and 'Now!'

Often these demands would be interspersed with swear words. He would request chicken, pasta, sometimes cold meats, but a particular type of German biscuit, covered with a slab of delicious dark chocolate, was one of his real cravings. 'If I took two [biscuits] up he would ask for three. Sometimes I would just give him two and that would be okay. But sometimes he would come down and he would search the cupboard looking for the packs of biscuits. We would have to be very creative in our hiding, but [if he found them] he would go through the whole packet.'

After these initial few days of marked stupor, Jamie would lighten a little. He would start moving around a little, playing music on a speaker in his bedroom. But his behaviour would be far from normal. His father describes him getting frustrated, and sometimes throwing the speaker across the room, trying to turn it off. Oriele recalls Jamie using the chocolate biscuits to mark off the days of his episodes on his bedroom wall, like a prisoner in Alcatraz marking the passing of his sentence. The marks are still evident above Jamie's bed. There are other examples of very bizarre behaviour that Jamie reports. I am not sure how much of it Jamie actually recalls, or is simply as a result of repeated telling by his parents, passing into family folklore.

On one occasion his mother came in with some food and switched the light of his bedroom on. 'So I said: "Mum, turn the effing lights off, you bitch," and then my mum said: "Jamie, that's not very nice!" So I said: "Please turn the effing lights off!" Clearly something in my brain was connecting, that it would be more polite to say "please", but I was quite childish.'

On another occasion, Jamie had been thrashing around and had knocked some of the stuffing out of his pillow. His mother came in and asked why his bedroom floor was covered in stuffing. 'I sat bolt upright and I said: "I've been shitting feathers!"' Jamie chuckles. 'We can look back now and just laugh because it is weird, it is just funny.'

What was far from funny were the effects on his education. In the year after his diagnosis, Jamie's events were happening on an almost monthly basis. 'My school attendance was down at 60 per cent. In a way I was fortunate, because a fair few of my episodes had happened during school holidays, so I didn't miss too much school.' I would shudder to think what Jamie's idea of really missing too much school is. But, despite his repeated prolonged absences, he managed to perform startlingly well in his exams.

'I got eight A*s, one A, and one B,' he tells me, trying to suppress his pride. I tease him that he is obviously an underachiever. 'I was very, very happy,' he beams. 'The school actually published a photo of me receiving my results. The shock on my face, the pure delight of getting my results. It reminded me that I was okay. And also how lucky I am that I was awake for my exams.' Jamie was also awarded a prize for his exam achievements in the face of his illness, but Jok interrupts: 'You missed collecting your prize. You were asleep at the time.'

The impact on his love life has not been insignificant either. In fact, his KLS precipitated the first contact between his and Megan's parents. Megan takes up the story.

'We hadn't been going out for very long, but I kind of knew he had this [condition]. We had gone to the cinema. I remember him being so rude towards me. I remember thinking, *He is not very nice. I don't think I like him any more*,' she laughs. As the film started, Megan saw Jamie fall asleep, reinforcing in her mind his impoliteness.

> When we came out, he just began walking and I didn't know where he was walking to. I didn't know London very well. I nearly did just walk away from him. But then I started thinking, *Could this be an episode? This is not him.* He took my phone away, so I couldn't try to call anyone and say he is not right. This lady was tutting at us because I had his arm over my shoulder to kind of hold him up. He looked like he was on drugs.

Now in full episode, Megan, a slight girl, had to try to get Jamie home. 'He's a big lad; it was quite a challenge.' Jamie is indeed a big lad. Even at the age of seventeen, he must be at least twelve stone of muscle. Eventually, Megan managed to get him onto a bus and succeeded in retrieving her phone from him. But even then, he refused to let her call his parents. In desperation, she called her own parents, who then spoke to Jok and Oriele for the first time. 'The first contact between our parents! It was over an episode,' Jamie laughs awkwardly.

Jok remembers coming home to find Megan and Jamie there already, Jamie fast asleep, and Megan sitting on the sofa, crying. 'We were familiar with it by then, we kind of knew what this was. It was obviously more distressing for Megan, who had never seen this. But we had a nice chat while Jamie slept, and we hadn't really met Megan properly till then.' But, as Megan went to leave, she went upstairs to say goodbye to Jamie. 'I said goodbye, and he sat up and told me to eff off!'

I am amazed at Megan's staying power. I suspect most teenagers would have disappeared at this point, but over a year later they are still together and going strong. If anything, dealing with the adversity of Jamie's KLS must have meant their relationship has matured much more quickly than it otherwise would have done.

Over time, Jamie's symptoms have altered a little. It is now three years since that fateful school skiing trip. His episodes are now less dominated by profound sleepiness. 'I just feel completely derealised,' he comments, describing that dreamlike state, the detachment from the world around him that is so typical of KLS. 'I just sit in front of the television. A big thing to keep me sane while in an episode is watching films I've seen before, or TV shows where I can predict what's going to happen. So quiz shows, for example. If I can think and answer [the question] in my head and then it pops up on the screen, for me that's reassuring. It keeps me grounded, connected to the world around me.'

After his school exams, Jamie went for six or seven months without an episode.

> There was always something in the back of my head [saying] that it was going to come back, it was going to hit me. Then it came back in October of 2017. It has been tricky, the past few months. I had one half term where I had a week on, a week off, a week on, a week off, and that for me was really difficult because trying to play rugby, trying to get to fitness, I feel I put a lot of weight on when I am in an episode. Also catching up with schoolwork – I kind of catch up and then I drop behind again.

The ongoing episodes beg the question why I haven't successfully treated Jamie. The short answer is that there is no successful treatment, or at least nothing that works consistently. Numerous drugs have been tried in KLS – stimulant drugs during a bout, and a range of pharmacological agents, mostly anti-epileptic drugs, on an ongoing basis in an attempt to reduce the frequency and severity of episodes. But what is clear is that no drug works particularly well. And because of the nature of KLS, clearly demonstrating that any treatment works is problematic. The bouts of KLS come and go, without a clear regular pattern, and so knowing if the KLS is responding to any particular treatment, unless the effects are dramatic, is difficult. KLS often gets better on its own, burning itself out after ten to fifteen years, if you are lucky, and so any improvement may simply be the normal behaviour of the condition rather than an effect of treatment.

The only way to answer the question as to the effectiveness of any particular drug is a randomised controlled trial, where patients are randomised to one treatment or another, so that the two groups can be compared. For this, however, you need reasonably sized groups of patients. The rarity of KLS, perhaps as rare as one per million of the population, i.e. as few as seventy people in the UK, or 300 people in the US, means that ever recruiting enough patients to any trial is nigh-on impossible. So, while there are several case reports of drugs appearing to have an effect in one or two individuals, the jury is still out.

Perhaps the strongest evidence is for two treatments. The first, as a preventative, is lithium, a drug more widely used for bipolar disorder. A systematic assessment has suggested it has some benefits. But it also has several side effects, some life-threatening, and requires very regular blood tests to ensure that the dose is not at toxic levels. Unsurprisingly, this treatment has never really attracted Jamie.

'I have always wanted to live my life as normally as possible, aside from KLS,' Jamie says,

> and so the treatment and the thought of having to go to a hospital regularly almost on a weekly basis for a check-up

> is not something that appeals to me. The treatment is not fully proven, is not conclusive. I just don't want to disrupt my daily life because I enjoy it while I can. I will have three, five, six days out in an episode, but once I'm back, I just want to live my life. I get back to school as quickly as I can, catch up on my work, I try to play rugby, I try to meet all the friends and my girlfriend. So I just try to live my life.

The other treatment is to bring people out of very long episodes, and involves giving patients high doses of intravenous steroids, and has been reported as being helpful in some people. But, given the duration of Jamie's episodes, it is not appropriate for him, and is equally not without side effects. So, for the moment, Jamie is not on treatment. Simply understanding his condition, liaising with his school, educating his teachers and friends about the syndrome, is about the limit of what modern medicine can currently do.

* * *

So, we are left with a syndrome rather than a disease, without any clear diagnostic tests and without any clear therapeutic options. The final issue I have been skirting around, however, is what we think the cause is. Are there any clues out there as to what might be going on?

The truth is there are very few. As I have already mentioned, KLS seems more common in Ashkenazi Jews than in other populations, raising the possibility of genetic factors at play. In fact, there have been a few reported cases where more than one member of a family has the condition, certainly implying but not confirming a genetic predisposition.

And what about which area of the brain is involved in KLS?

Even this is uncertain. As with many sleep disorders like narcolepsy, the hypothalamus has been implicated, but there is little evidence of hypothalamic abnormalities in most patients with KLS. There are rare reports of patients with crack cocaine-induced strokes affecting the thalamus, causing sleepiness, hypersexuality and aggression, and this would obviously fit with research scans showing abnormal activity in this region of the brain during episodes of KLS. Postmortem studies of the brains of people with KLS are rare, as KLS is not a fatal condition, but there have been four patients who have died of a KLS-like condition whose brains have been closely studied. In three, there was evidence of inflammation in the thalamus and hypothalamus. It should be stressed, however, that these patients are probably not representative of 'typical' KLS, but they certainly raise the possibility that an immune or inflammatory process might underlie this condition.

Like other neurological conditions thought to be caused by the immune system attacking itself, e.g. narcolepsy, many patients or families report a viral infection prior to the onset of KLS. However, one always has to take reports in hindsight with a pinch of salt. Ask anyone if they remember having a cough or cold over the past few weeks and they will probably say yes, particularly if they live in London in the winter months and use public transport! But the young age of onset and the intermittent nature of KLS (a bit like multiple sclerosis, another autoimmune neurological disorder) for some researchers also implies that the cause is our immune system gone awry. It certainly looks as though viral infections can trigger people to go into bouts. Recent studies have shown bouts to be more common in winter and spring, correlating with the incidence of upper respiratory tract infections.

For me, one potential cause is what we neurologists term a

'channelopathy'. Neurones, the nerve cells that constitute the major functional part of the nervous system, utilise a complex system of pumps and channels across their outer envelope, the cell membrane. Imagine the cistern of your toilet, into which water is slowly pumped under pressure. When you turn the flush handle, a valve opens and a torrent of water floods into the pan. Similarly, cells generate a difference in ions, electrically charged salts, between the inside and outside of the neurone. Like the water pump filling the cistern, molecular pumps use energy to shift ions from one side of the cell membrane to the other. But when electrical impulses are conducted along neurones, the flush handle is turned. Gates in the cell membrane open, allowing ions to flood through the cell membrane, like water into the toilet pan. And because these ions carry an electrical charge, there is a sudden change in electrical charge across the cell membrane. This torrent of ions is the basis for the impulses that underlie not only our nervous systems, but also our hearts and skeletal muscle. And so these molecular gates, also known as ion channels, are fundamental to our biology, to our ability to think, move, indeed live. But these ion channels are far from simple.

Structurally complex, they are made of multiple subunits, encoded by multiple genes. Some are triggered to open by electrical discharges nearby, others by the binding of chemicals like neurotransmitters. And, to add to the complexity, the composition of these channels changes throughout our lives. Different subunits are generated by our genes at different ages, so that some ion channels may alter their properties as we move from infancy to childhood to adolescence and so on. This probably explains why certain neurological disorders known to be related to ion channel problems start at particular times in our life and then disappear as we age.

Over the past few decades, with advances in genetics and molecular techniques, our understanding of many common and rare neurological disorders has changed beyond recognition. We now know that many of the conditions we see routinely in our outpatient clinics or in the emergency department are essentially disorders of these ion channels – conditions like migraine or epilepsy (the genetics of one group of ion channels in epilepsy was a focus of my doctoral thesis). Sometimes these disorders – channelopathies – result from mutations in the genes that code these channel subunits, like familial forms of epilepsy, migraine, or more unusual conditions like episodic ataxia, where sufferers experience attacks of imbalance and incoordination lasting minutes to days. Occasionally, channelopathies arise from antibodies directed against these channels, although these do not tend to be quite so episodic or intermittent.

KLS, therefore, has multiple features of a genetic channelopathy. It often comes on in adolescence, like many forms of epilepsy, migraine or other types of these disorders, and is intermittent, often triggered by tiredness, stress, alcohol or illness. Many of the drugs that have been reported as being helpful in KLS, such as lithium and anti-epileptic drugs, act on these ion channels. At present, there are multiple groups, including my colleagues and me, who are searching for genetic mutations that may underlie KLS. Many of my patients have volunteered to have their whole genome sequenced, to ascertain their entire genetic code. But we are searching for a needle in a haystack, potentially a single abnormality in a 3-billion-letter sequence. And the change may be different in each person with KLS, or indeed it may require multiple changes in multiple areas of the genome.

Tantalisingly, one of the leading research teams, headed by

Emmanuel Mignot of Stanford University, the man responsible for identifying the cause of narcolepsy, has found a signal in an area of the genetic code near to a gene that has been implicated in bipolar disorder, another disorder associated with abnormal ion channel function, and with some shared features with KLS. There are also some features on studies of the constituents in the cerebrospinal fluid, removed by lumbar puncture, to suggest that there may be subtle evidence of inflammation in the brain after all. This links well to recent theories of depression and other psychiatric disorders being caused by brain inflammation, but, given the complexities of KLS, and its rarity, the challenges are huge.

Then, when you thought it couldn't get any more difficult, it may be that KLS is more than one disorder. The range in the severity and duration of episodes is very wide, and my own view is that we sometimes misdiagnose KLS. I and my colleagues have seen a number of patients in whom their bouts are short, two days or a little more. This is in stark contrast to some people with KLS who will report bouts lasting several months. According to the diagnostic criteria, these patients with very short bouts meet the definition of KLS, but careful assessment, obsessional taking of the history, reveals features in some that hint at an alternative explanation.

A very prominent headache before the attack, severe nausea or imbalance, some tingling starting in the face, then spreading to the arm – these are all features of migraine.

For most people, migraine simply means severe headache. Many people will have experienced a migraine, and it is not a particularly pleasant experience. My migraines seem to have a particular predilection for when I am sat in my clinic room, usually with a waiting room full of patients. I suddenly begin to notice a flickering in the periphery of my visual field, a little flashing or shimmering. Over the course of a few minutes, this effect, like the heat haze above an asphalt road on a sunny day, spreads and moves more centrally. As I peer at my computer screen, flashing with patients waiting more than thirty minutes, I find it difficult to focus on the print. If I am lucky, my colleague will be running a migraine clinic next door. I know she is more organised than I am, and will often have a supply of medication for her own migraines. I will pinch a couple of ibuprofen tablets and, within a short while, my vision will return to normal. I will be left with only a slight heaviness in the head.

Like me, many people with migraine will experience these visual phenomena – sometimes shimmering, sometimes zig-zag lines, sometimes flickering – slowly marching across their eyes. And so migraine is not just a headache. For many, migraine also involves some dysfunction of the nervous system. In the case of a visual aura, as these phenomena are called, we know that there is a wave of abnormal electrical activity passing slowly over the cerebral cortex in the occipital region, the back of the brain, where vision is processed. But auras are not only visual. They can also involve other neurological functions.

Patients with migraine will often experience word-finding difficulties, or tingling in their face or limbs, as the dysfunction of the aura spreads over their verbal or sensory areas. Many of us neurologists will have had to run down to the emergency department for an urgent stroke patient to find that the pattern of weakness is in keeping with hemiplegic migraine, where the dysfunction has affected the movement part of the brain. A few neurologists might even admit to having given clot-busting medication to patients who in hindsight were simply having a migraine attack! And for some people with migraine, the aura consists of symptoms associated with the brainstem – dizziness, imbalance, incoordination. But, as we know, the brainstem is also fundamental to the regulation of sleep and the maintenance of consciousness. In extreme cases, people have been reported going into a coma during migraine attacks.

So, migraine is not purely a headache as some people believe it to be, and, in some of the patients I see, there are strong pointers that they have an unusual form of migraine rather than classical KLS. In a few, simple treatment with an anti-migraine drug has led to startling improvements. And thus my practice is always to find an alternative explanation to KLS, on the basis that we know that treatment of KLS is problematic to say the least, whereas treatment of some of the alternatives is much more straightforward.

* * *

Since Jamie's diagnosis, the entire family has become very involved with the principal charity in the UK to support KLS patients and their families. Jok has rolled up his sleeves and has become a trustee for the charity, as well as actively helping to organise various projects to publicise Kleine–Levin syndrome. Obviously, those with KLS are incredibly keen to help with research into the understanding of their syndrome, and the treatment of it, but it is a small charity, without significant financial support, and the focus remains on getting people diagnosed as quickly as possible. Most, if not almost all, of my own patients and others in this support group have taken a long time to get a diagnosis of KLS, and their parents have often struggled physically and psychologically while trying to support their kids at a delicate and formative stage in their lives. The parents invariably say that, at the time of onset of symptoms in their child, the single most invaluable thing would have been knowing what it was that was happening to their child, and perhaps talking to other families who had gone or were going through the same torment.

I and one of my paediatric colleagues act as medical advisers to this charity, and over the years I have attended several of their annual meetings. When I stand there discussing the latest research, or lack thereof, I am struck by the range of this condition. Some of those patients present have attacks once a year, maybe even less. Several have gone for several years without attacks, while a few have gone for a decade or more before having a further attack out of the blue. Others have attacks every month, perhaps even more frequently, and in a few of these patients, KLS defines their lives. They cannot maintain jobs, relationships, social lives, or an education. For them, KLS becomes who they are, a debilitating condition that overrides anything else. And for their parents, KLS often takes over their lives as well. The unpredictability of KLS bouts and the complete vulnerability that ensues leaves parents on tenterhooks, waiting for the next phone call from school, from friends, or from their child's workplace, to rush home to look after their incapacitated offspring.

Within the spectrum of severity, Jamie's KLS sits somewhere in the middle. Not the worst of any of my patients, but not the mildest either. I am, however, amazed by how little Jamie allows his KLS to define him. In part, this is due to the fantastic support network he has around him. His parents, Jok and Oriele, are constantly there for him, but are practical people who focus on facilitating Jamie to lead the life that he would lead without KLS. His girlfriend Megan is obviously a rock – I am not sure many seventeen-year-old girls would have the maturity to cope with a partner with KLS and all that it entails. Jamie says of his family: 'My family and Megan are really accepting – "He is in an episode, he is doing things, saying things he doesn't mean, and he wouldn't normally do." I can't imagine how tricky it is for them to watch me go through these things and know that as much as they try, as much as they care, there is not a lot they can do. They just kind of have to sit back and let it pass.' Jamie also has a close circle of friends who look out for him.

Whenever I see Jamie, though, what is most evident is his strength of character, his resilience, and his desire to get on with his life. I ask him if he, like other people I look after, has a personal relationship with his illness. He responds: 'I feel it is part of me, but because it is a temporary thing, because it comes every now and then, I don't feel any kind of personal relationship with it. It is just there, it lingers, but I push it to the back of my mind and forget about it. There are days when I don't even think about it.'

Jamie lives in the hope that one day a bout of his KLS will be his last. 'It will hopefully subside. This comfort is driving me, because hopefully, one day, I will look back and say, "See what I've done despite KLS."'

And as I sit around a table with Jamie and his family, I have no doubt that Jamie will go on to achieve many things.

I look over to Oriele and Jok and think to myself that if Jamie were my son, I would feel rather proud.

13

INCEPTION

I love Christian. I love it when I see his name on my clinic list for the day. He will always make me laugh. Sometimes, my sleep clinics are rather emotionally draining, since many people are debilitated by their sleep problems, and have associated psychological and social problems. Christian, however, is like a breath of fresh air.

Please don't get the wrong idea, though. Christian has severe narcolepsy and cataplexy, to the extent that he cannot maintain employment, and finds it difficult to lead a normal life. But he has something about him, a slightly off-beat, occasionally surreal view of the world, and a lovely turn of phrase that is a pleasure to listen to. As he sits back in the armchair in my clinic room, casually dressed, with a neatly trimmed beard and cropped hair, he tells me of his latest narcolepsy experiences with a degree of insight, humour and analysis that I rarely encounter.

Intelligent and eloquent, he has been living with narcolepsy since his teenage years. And my impression is that he likes to talk about his narcolepsy with me because he feels that no one around him understands this aspect of his life. 'Apart from close friends obviously, you're one of the only people that I can actually sit down with and tell about these experiences,' he says. 'And you're not looking at me like I'm just insane, you know?' I laugh. 'I mean you might *think* I'm insane, but you don't look at me in that way!'

Christian experiences almost all the features of narcolepsy, which started when he was a teenager but, like many people with this disorder, he was only diagnosed after many years, at the age of thirty-two. He is now forty years old, and he has the profound sleepiness that plagues all with this condition. It is this that is the biggest issue for him, the one that prevents him from working. 'I've fallen asleep – and had some really funny looks – at White Hart Lane, watching the Spurs v Man City game,' he tells me. 'It was a packed ground – we're talking 30,000–40,000 people – shouting, singing and hollering. It wasn't quiet.' I am not known for my love of football, and I am not sure I would mind falling asleep at a match, but of more concern is some of the other situations in which Christian has found himself dozing off.

He previously worked for a company that made medical instruments. Part of his job would involve him driving a fork-lift truck, loading large containers onto delivery lorries. He would often park somewhere out of the way, out of view of his colleagues: 'I would rest on the steering wheel and I'd literally go to sleep. It could be for a minute, it could be for ten minutes.' He was not always able to pre-empt his naps, however. 'I've driven into them [the lorries] before. I've driven along, picked up a pallet. And then I've woken up with a cold feeling – a jolt, sudden reality thing' – he claps his hands loudly, miming the crash – 'as soon as I've hit the container. Bang. Oh, Christ! Luckily it wasn't a person or anything valuable.'

Equally worrying is that he has occasionally drifted off while walking down the street. 'I remember not being able to control my eyelids and sort of drifting off, falling asleep. But when I got to the road, and a car – voom!' – he simulates a car driving past him very close – 'it suddenly snapped me out of it. I could have walked out in front of that car. A cold reality hit me.'

Christian also used to work on the hovercraft that ran between Dover and Calais, hurtling across the English Channel. I recall taking such a Channel crossing as a child. It was a relatively calm day, but nevertheless I vividly remember the noise and being thrown up and down with every wave, worse than the most turbulent flight I have ever been on. But for Christian, the discomfort of the hovercraft was no problem. 'I used to just go to sleep. Like pretty much every job I've ever had.'

Like Phil and Adrian, Christian also experiences cataplexy. For him, the trigger seems to be mainly laughter. 'I might watch *Fawlty Towers*, and John Cleese'll do something funny and I think it's the most hilarious thing I've ever seen in my life. And that triggers the cataplexy, [affecting] basically all my muscles. My cheeks and my facial muscles go a bit funny. It's like they're wobbling. I lose control of my arms and my legs.'

A few years ago, I and a couple of my colleagues tried to record Christian's cataplexy in an MRI scanner, by playing one of his reliable triggers on a screen while he was inside the machine over and over – the scene from *The Office* when David Brent shows off his dancing skills to his colleagues, in a buttock-clenchingly awful display of self-delusion and lack of insight. It was a resounding failure, not because we could not trigger the cataplexy, but because of the technical difficulties of combining the scan with a recording of his brainwaves.

And, like Evelyn, Christian also has hypnagogic hallucinations. He will often see dark figures in the room as he drifts off to sleep, but he also has had even more bizarre hallucinations. He has occasional memories of floating above himself, like an out-of-body experience. Christian would also experience something like the descriptions of the incubus or succubus in Chapter 9. 'I'm not a religious person, but I believed that I was being visited by a succubus, a demon, of a night-time. It was a hallucination of smoke. It wasn't a demon in a person form or anything. It was like a thick smoke that would come to me.' He hesitates. 'In my mind this thing was coming to me, having sex with me, and then leaving. And it would do this every night. I could see the smoke even though I was asleep with my eyes open. I could grab it. Not something sitting on top of me as such, but something sort of entering me – without sounding crude – in a sexual way.'

He adds: 'I was even ready to go and see a priest, actually just to have a chat with him. But I came to see you first!' he smiles. I tell him I am a poor substitute for a priest.

I presume the basis of Christian's narcolepsy, as in Phil and Adrian, is damage to that small area deep in the brain called the lateral hypothalamus. I have never checked his levels of hypocretin, the chemical produced by this region, in his spinal fluid, but I have no doubt that it would be low or even completely absent. The destruction of this tiny nucleus at the very centre of the circuitry that controls our sleep and regulates dreaming is responsible for all his symptoms. The switch that stops him suddenly drifting off to sleep, or keeps him from entering into REM sleep, has gone awry. For most people, our sleep cycles through REM sleep four or five times every night, and we may perhaps remember a dream if we wake in the morning out of REM sleep. But Christian flicks rapidly in and out of sleep, resulting in his sleep attacks at highly inappropriate times, such as behind the wheel of his fork-lift truck or in the baying crowd of a football game. And he goes directly into REM sleep from wake, resulting in these strange and scary hallucinations as he lies in bed.

In addition, this neurological damage results in his REM sleep being unstable, so he judders constantly between REM sleep and wake at night. And it is in the blurred margins between REM sleep and full consciousness throughout his nocturnal hours that he recalls his dreams. Not a fragment of a memory or a story, dimly recalled on waking, but constant, vivid, real, recurring dreams, so intense and colourful that they are sometimes difficult to distinguish from real life. I am fascinated by his descriptions.

Christian dreams of the same four people almost every night. 'Real people that I went to primary and secondary school with,' he tells me. 'These four people that I haven't seen for maybe twenty years, but they're in my dreams every night.' He laughs slightly sheepishly. 'I'm having a bit of a love affair with one of the girls. Actually two of them!' He feels the need to clarify: 'This sounds really dodgy when I say a love affair. The people that I'm actually seeing in my dreams are kids. I'm not having a love affair where I'm an adult and they're children. I'm the same as them. I'm the child as I was in the past.'

I ask if he is simply reliving old memories, but he is clear that he is living new experiences, almost a new life. 'One of them used to be a girlfriend of mine, and another one was just a classmate that I quite fancied. So I'm having this really weird love affair that's actually not happening in reality in any way, shape or form – I mean I haven't seen them for twenty years! I have seen pictures of them on Facebook, so I do know what they look like as adults, but I don't dream of them like that because I've never interacted with them.' He leads a separate life at night – a social and romantic one – but he is clear that his relationships do not have a sexual aspect to them. In his dreams, it is a time of innocence, a period of happiness in his life, without worries or concerns.

As well as the content, it is also the intensity of his dreams, and their persistence, despite waking, that is equally arresting. He finds it difficult to describe, and illustrates it with an example.

> Say in the dream I'm walking down the road and I meet Fred Bloggs. In the dream I'm saying, 'Hello, Fred, how's it going?' I'm seeing it from my point of view. As soon as I wake up, as soon as my eyes are open, the dream is still going on. But from that point on, I've gone from just seeing Fred Bloggs and saying hello to it suddenly being narrated to me. It will suddenly turn into a voice that is reading the book to me. So I'm talking to Fred Bloggs, [then suddenly the dream becomes] 'Christian's talking to Fred Bloggs. He shakes his hand.'

My interpretation of this is that aspects of his dream, but not the full dream, are leaching into wakefulness. I ask him what happens when he drifts back to sleep. He says that the dream then simply continues, without the narrative voice. He finds himself back in the street, talking to Fred. 'The dream just continues seamlessly,' he replies.

* * *

As I watch my younger daughter in her bed, I am always amazed by – and envious of – the speed with which she falls asleep. One moment, she is pestering yet again for a puppy, a final thought before sleep; the next moment, her eyes are closed, her breathing slows, and she is out like a light. It is as if she has fallen off a cliff, deep into the ocean of sleep, unaware of the external world. The transition is sharp, immediate, sudden. I am sure that she never gives sleep a second thought. It is instinctive, primordial, a biological act, like eating or drinking, something that just happens. She does not consider what sleep is for, other than not feeling sleepy, and has no awareness of herself asleep – with one exception. She is aware that she has dreamt, or has had a nightmare. For most of us, at least those of us for whom sleep is not an issue, dreaming is the only aspect of sleep that percolates into our consciousness. Apart from the physical act of waking up, it is the only evidence that we have actually been asleep. And it has always been like this. Dreaming is an intrinsic part of the human experience, and maybe the experience of other mammals, although of course we cannot ask them. Humans have been puzzling over the meaning or significance of our dreams for time immemorial, and dreams have some part to play in almost every religion.

One of the earliest written texts still in existence is the *Egyptian Dream Book*, a papyrus document now housed in the British Museum in London, only forty minutes' brisk walk from Guy's Hospital. It is too fragile, too delicate, to remain on display. Exposure to even the low levels of light in a display case risk the pages crumbling to nothing. Dating back to about 1220 BC, it details 108 dreams, categorising them into good and bad, interpreting what these dreams predict. I have been fortunate enough to see these ancient texts for myself.

After being ushered into the ancient Egypt and Sudan study room, four frames of glass are brought out and gently placed on the oak desk in front of me. Sandwiched between the glass plates are the papyri themselves, some mere fragments floating, disconnected, while one papyrus in particular is surprisingly large and intact, about 2ft wide by 1ft high. I can see the feathering of the script in black and red ink, where the scribe's hand has risen off the page. It is astounding to see something written over 3,000 years ago with ink on papyrus, and to see the marks of its author. To my eye, it actually looks like the work of more than one person, with different-sized symbols. The main text is precise, regularly spaced, but in areas of the documents, a different hand has added to it; larger, more florid, less controlled, like comments in a margin. And, although hieroglyphics is not my strong point, I can make out what I take to be a list, each sentence or phrase marked by a red symbol that I presume to be a bed. I wonder if each item is one of the dreams and their interpretation.

The ancient Egyptians believed that their gods could show themselves to us in dreams, and that dreams could also serve as a window into the netherworld. Similarly, Joseph, in the Book of Genesis, predicted seven years of feast followed by seven years of famine by interpreting the dreams of Pharaoh. In Kabbalah, a branch of Jewish mysticism, it is believed that, in sleep, fifty-nine out of the sixty parts of our soul leave us, with one-sixtieth left behind to sustain us – one-sixtieth being seen as the smallest detectable amount of anything. While our souls enter the spiritual realm to be nourished, glimpses of events coming our way trickle down into our bodies, taking the form of dreams.

Few people would now consider dreaming to be a prediction of the future, but the answer to the question, 'Why do we dream?' is a resounding, 'I don't know.' If you think about it, this is staggering. Something that is fundamental to the human experience, that we all do on a nightly basis, much like eating and drinking, remains a mystery. The concept of asking 'Why do we eat?' and being answered 'I don't know' is totally preposterous.

We consider REM sleep and dreaming to be the same thing, but as I have already said in Chapters 2 and 3, we know that this is incorrect. We know that if you wake someone up in non-REM sleep, they will often describe dreamlike phenomena, but it seems that REM sleep is most closely correlated with those dreams of narrative structure, of a plot evolving in your mind, a story that unfolds. But REM sleep and this type of dreaming are clearly not precisely the same thing. REM sleep is defined as the state of sleep during which your body is paralysed but your brain is very active. The EEG, the electrical signals that give us a limited insight into brain activity, looks remarkably similar to the awake brain, but the only muscles moving are those that allow us to breathe, and those that move our eyes. REM sleep is also a feature of brain activity that starts in the womb. In fact, as we enter the third trimester of our development, almost all our time is spent in that brain state, at least in premature babies born at this stage. And, even after birth, a third of each 24-hour period is spent in REM sleep.

As Meir Kryger, professor of sleep medicine at Yale, told me when we were discussing this: 'What in the world can babies be dreaming of?' The answer is probably nothing, or at least nothing like what we know of as dreams. It seems likely that our experience of dreaming, in the way we understand dreams, only really begins at an age when the brain is developed enough to represent subjective experience in a narrative way. What I mean by this is that our brains can only understand our experiences of life, and put them together in a story-like way, when they reach a particular stage of maturity and organisation. And the age at which our dreams coalesce into what we recognise to be dreams is probably somewhere around five years old, although it is difficult to ascertain. It is difficult enough to understand why your two-year-old is having a tantrum, never mind interrogating them as to their dream experiences, and certainly young children will wake up and say they have had a bad dream.

But, as Kryger alludes to, how can a foetus at twenty-eight weeks old dream of life experiences, if it has no experience of anything outside the womb? Allan Hobson, professor of psychiatry and sleep medicine at Harvard, has proposed that this 'dreamless REM sleep' is fundamental to the development of consciousness, what he has termed 'protoconsciousness'. Our behaviours in infancy and childhood, which start out as unconscious automatic actions – crying for food, suckling, reaching for objects without any self-awareness or volition – gradually become what we perceive to be conscious, through rehearsal and 'practice' in a virtual space. He argues that REM sleep is a driver of the emergence of secondary consciousness – that is the degree of consciousness that differentiates us from other species, namely volition, self-awareness, reasoning, insight, abstract thinking – from primary consciousness, simple perception and emotion that other mammals experience. Hobson says, 'I think protoconsciousness is an unfortunate term which I used, now with some embarrassment. But I think that what I am trying to get across is very simple – that the brain development that leads to consciousness occurs very early, it occurs in pregnancy.'

However, while REM sleep is a brain state – measurable and detectable – dreaming is a subjective psychological state. At least we have some sort of correlate of dreaming in REM sleep, something that is easy to determine and study. So let's set dreaming aside for a while and concentrate on REM sleep. What do we know about its function? Well, if we spend very high proportions of our time in it as our brains develop, in the womb and in childhood, perhaps it has a role in promoting that development. But REM sleep persists throughout our lives, into old age, so this cannot be the whole story. Maybe it also has a role in the ongoing maintenance and reconstruction of our brain? Perhaps REM sleep allows us to practise or prepare for waking life, allowing us to sleep and fine-tune our brains for when we wake. Jouvet, he of the cats with lesioned brains that appeared to act out hunting or fighting in REM sleep, proposed that REM sleep allowed the rehearsal of instinctive behaviours crucial to survival. Others have pointed out that REM sleep in kittens drops off markedly as soon as they open their eyes, suggesting that REM sleep somehow prepares certain circuits in the brain for future functions or activities.

An alternative theory regarding the function of REM sleep is that it is fundamental to the regulation of mood and one's psychological state. While initial studies disrupting the REM sleep of human subjects showed that, after a few days, people began to experience significant problems with mood, other studies have since pointed to REM and non-REM sleep being equally important. To confound matters, a single night of preventing REM sleep in people who are depressed can cause marked *improvement* of mood, and older antidepressant drugs can completely abolish REM sleep. So this is not an entirely convincing theory either.

What about as a facilitator of learning and memory? Certainly, a common finding from animal studies is that REM sleep is prolonged after learning a new task, and that disruption of REM sleep also disrupts the learning process. The evidence from human studies is for the most part weak, however, and if there is such an effect, it must be relatively minor. So, maybe REM sleep is not only about storing new material, but also about maintaining what we already know. Perhaps REM sleep triggers circuits that encode memories or skills, reinforcing them and preventing us from forgetting them. Evidence certainly points towards REM sleep being fundamental to a range of aspects of learning and skills. Matthew Walker, one of the foremost researchers in this field, and his colleagues have shown that important abilities like recognising the emotions expressed by faces, maze navigation and creative thinking are all facilitated by REM sleep. Creativity and dreaming have long been linked. The examples of Mary Shelley dreaming up scenes from *Frankenstein*, Keith Richards and the opening bars to 'Satisfaction', Mendeleev conceiving of the periodic table and Paul McCartney composing 'Yesterday' are often invoked. So, maybe REM sleep has the power to coalesce multiple memories, disparate experiences, and integrate them into a work of genius.

One peculiarity of REM sleep is that, unlike every other moment of our lives, the mechanisms that regulate our body temperature fail us. While for the rest of the time our body temperature is kept absolutely stable, in REM, it drops. This is actually a highly dangerous state for us to be in. Even small fluctuations in temperature can result in our brains not working properly, or heart rhythm abnormalities. I will occasionally see patients in the intensive-care unit rendered comatose through hypothermia, or with brain damage resulting from overheating due to recreational drug use. Given that the loss of temperature control is so hazardous, it seems a high price to pay, especially as we enter this state multiple times every night. Evolution is no fool, and this implies that REM sleep, and this loss of our internal thermostat, must have a crucial function.

There is a safety cut-off, however: if we are cold, we will not enter into REM sleep. It seems that if we are not warm, the brain will not risk losing control of heat regulation. But there is a paradox here: if rats are prevented from having REM sleep, they soon lose the ability to regulate their body temperature, and soon die as a result. Hobson, when I speak to him on the telephone, likens this to his wife. 'My wife', he tells me, 'is a neurologist too. She always feels cold in a room that everyone else feels is hot. This is because she is always sleep-deprived; she works too hard!' So, if REM sleep is important for the regulation of our body temperature, how does this fit with us losing this control while we are in REM sleep? I ask Hobson what he thinks of this apparent paradox. 'When we go into REM sleep, we are fixing the system that regulates our body temperature. It's like when you take a car to the shop [mechanic], you run the car with the wheels off the ground. You stop it working normally to fix it.' So, maybe by taking the system that maintains us at a steady 37°C offline, it provides an opportunity to tweak the system, to maintain it and keep it in good running order.

* * *

For Christian, it is not just dreams of his childhood friends; his recollection of his dreams is staggering. I am lucky if I remember a dream every couple of weeks. I find that on waking I can clearly recall details of the dream, but over a few minutes, these intense memories dissipate like wisps of mist in a breeze. For Christian, the details live with him, as bright and colourful as his daytime experiences. 'There's always something special about my hands in my dreams. I can always shoot lasers out of my hands. I can describe exactly how it happens in every dream. It's always the same. I feel a pressure in my brain. I well up this ball of energy. I can feel it in my brain and I can feel this ball of energy getting bigger and bigger in my hand because I'm creating that ball of energy. Then I just fling it at people. I'm chucking balls of energy.' I tell him it all sounds very *Harry Potter*, which triggers off other memories in him.

'I walk through walls in my dreams, but when I walk through the wall it's difficult to get my head through.' He pauses, and thinks for a few seconds. 'I wonder if there's something going on in reality in my brain. I don't know if the brain swells a little bit. I don't know the science of that, but possibly in reality something's happening while I'm asleep? It's being transferred into my dream and I can't get my head through the wall?'

I suspect that he is describing the phenomenon of sensory information being integrated into dreams, as we discussed in

Chapter 3, where perhaps abnormal movements of the limbs generate the violent dreams characteristic of REM sleep behaviour disorder. 'Imagine if I had a spot, say, growing on my leg, but I didn't know it was there, I didn't notice it. That particular night I will almost certainly dream that there's something there, but in my dream it will be something like a mushroom growing out of my leg.'

Christian's dreams are not all sweetness and light, love affairs and superpowers, however. Some of his dreams are deeply unpleasant. 'A regular one is nuclear war. Always vivid. Horrible really. I've never been in a nuclear war obviously, I'm only going by video games, films, documentaries, that sort of thing.' He goes on to describe his regular experience of a nuclear holocaust at night, dreams, or rather nightmares, filled with horror, death and destruction. 'If, for example, we're in the dream now, right this second, I'll be sitting here talking to you and then the big mushroom cloud will go off, say, over in the distance. We start to panic or whatever. I can see the bombs coming down and I know where they're going to land. And the bombs never go off immediately. We'll leave the building trying to get to shelter somewhere. I've got to get underground somewhere. And then a bomb will go off right next to us. In reality you'd be vaporised in milliseconds, but in my dream, I have time to run away. Generally I don't die; I'm always trying to save people.'

* * *

Christian's regular recollection of his dreams is obviously a function of his unstable REM sleep, caused by his narcolepsy. As he shifts almost seamlessly between wake and REM and back again, his dreams regularly enter into his consciousness. The sheer intensity and quality of his dreams makes him wonder if there is some purpose to them, whether they are telling him something. But do they? Do dreams tell us something about us, our experiences, our desires, our personalities?

The science of dreaming has undoubtedly been hindered by the historical rift between mind and body (in this sense, 'body' includes the physical brain). The concept of Cartesian dualism defines the body and soul to be entirely separate, and this view has influenced the schism between neurology, the medicine of the nervous system, from psychiatry, the medicine of the mind. In the past few decades, this artificial division has been slowly whittled away, and few neurologists or psychiatrists today would consider this view correct. We neurologists are very familiar with brain tumours or autoimmune brain disorders causing frank psychotic symptoms such as hallucinations or delusions, or even more subtle 'psychiatric' symptoms, such as anxiety or depression. Equally, looking through the current psychiatric journals, they are filled with research correlating conditions like schizophrenia or bipolar disorder with genetic variants, changes in neurotransmitters and altered activity in various parts of the brain.

So it seems that the separation of mind and body, the mental and the physical, is slowly dying. In the world of dreams, however, all roads lead to Sigmund Freud. Freud's theories regarding the origins of dreams are familiar to almost everyone. The concept that our dreams represent unfulfilled wishes or desires is the central tenet of Freudian theory. But the overt meaning of our dreams represents a censored version of suppressed emotions or desires, requiring interpretation. Our unconscious desires are repressed, from others but also ourselves, and our dreams are covert manifestations of these hidden wants. Rather than needing the *Egyptian Dream Book* to translate our dream content into predictions of the future, we need a psychoanalyst to tell us what they mean with regard to the dark depths of our mind. In some respects, Freud was a bridge between the two sides of Cartesian dualism. A neurologist by training, he quickly moved into the realm of psychiatry and developed his famous theories of psychoanalysis. Despite being in the era before the EEG and brain imaging, and not even knowing about REM sleep, he firmly placed dreaming in the physical world, originating from the brain. But his theories on dream interpretation were untestable, unprovable, and definitely had a flavour of the soul about them. Our inner spirits being tortured by our latent dark desires on a nightly basis, a manifestation of our Oedipal complex, or repressed sexual scenarios of childhood.

Obviously, we can now clearly demonstrate that dreams originate in the brain. It is not only that dreams demonstrably arise in a measurable electrical brain state – REM sleep. From functional imaging studies, techniques that allow us to peer into the brain and understand the activity of various areas within it at certain points in time, we know that during REM sleep, various parts of our brains are extremely active. Simply thinking about your own dreams might allow you to predict which areas these might be. The hottest areas are those involved with emotion, movement, vision and autobiographical memory – the limbic system, motor cortex, visuospatial areas and hippocampus respectively. And, equally unsurprisingly, the major area that is quietened by REM sleep is the prefrontal cortex, the region of the brain responsible for rational thought and complex planning.

But could it be that dreams are just an 'epiphenomenon' of REM sleep? Are the mental processes of dreaming simply a result of the random activation of these areas of the brain in REM sleep? Consider the simple act of taking a breath and exhaling. As you do so on a cold day, the steam of your exhalation is testament to the water that we lose with every breath. For an average-sized man, some 400 ml of water is lost in this way every day. But getting rid of water is obviously not the function of respiration. The water loss is simply a consequence of air passing over our moist airways, a by-product of the need to inhale oxygen and exhale carbon dioxide – an epiphenomenon. So, is dreaming similarly just a by-product of REM sleep, meaningless garbage that is simply a consequence of REM sleep performing some housekeeping role in the brain? On a personal level, it is hard to reconcile this view with our own experiences of sleep – dreams in which we see people we know, interact with realistic worlds, experience strong emotions. Even trying to be as dispassionate and scientific as I can, I find it difficult to understand how our dreams could represent the irrelevant dregs of an electrical process that has an entirely separate and unrelated purpose. And when I look at some of my patients, it is equally hard to comprehend.

I think of a young man I saw recently, a refugee from the island of Sri Lanka. Born into a Tamil family, his early life was torn apart by the civil war, with intermittent periods of arrest, harassment and exposure to violence. Despite leaving Sri Lanka as a teenager, he is plagued by vivid nightmares every night or two, waking up screaming after the intense replaying of his traumatic experiences, even fifteen years later. These nightmares are a common feature of post-traumatic stress disorder (PTSD), associated with daytime flashbacks and abnormal responses to reminders of experienced trauma. So, how could these recurrent nightmares, clearly linked to daytime experiences, simply be an epiphenomenon? It seems rather unlikely.

In fact, one of the proposed functions of dreaming in REM sleep is what Matthew Walker terms 'overnight therapy'. In wakefulness, our brains are flooded with various neurotransmitters, but as we enter sleep, this changes. In non-REM sleep, levels of acetylcholine, serotonin and noradrenaline (norepinephrine) fall away. But, as we enter into REM sleep, this once again changes. Noradrenaline levels in particular fall to their lowest, but acetylcholine is boosted to higher levels than seen in wakefulness. So, on a chemical level, the brain in REM sleep is active, but without noradrenaline, which, like its related hormone adrenaline, is fundamental to the 'fright-fight-flight' response. This chemical may also have a role in intensifying memories associated with strong emotions; if you look back on your life, your strongest memories are almost always those experiences associated with extreme joy, fear or excitement, and this is certainly the case for people with PTSD. However, elegant research has suggested that REM sleep may provide a mechanism to consolidate these memories, to uncouple them from their emotional context – in other words, to lessen the intensity of emotion associated with your experiences. Think of it from an evolutionary perspective: if your abiding memory of being bitten by a snake is the fear and pain, then the next time you see a snake you may simply freeze with fear, like some people with PTSD do when something reminds them of their traumatic experiences. It is important to remember that snakes bite – and the pain and fear of being bitten certainly helps – but also to be able to maintain rational thinking the next time you see one.

So, perhaps REM is some form of psychological therapy, allowing us to rid ourselves of the emotional burden of some of our experiences. And maybe that explains why people with PTSD experience these recurrent nightmares. The traumatic experiences have generated such strong emotions, and the levels of noradrenaline are not fully suppressed, that these dreams or nightmares result in full awakenings. Each time the nightmare occurs, it is never completed, the emotions associated with that memory are not dampened down. Like a scratched record, the needle jumps out of the groove at the same point, and the song is never completed. The brain tries again and again and again to cleanse your memories of the fear, of the emotional trauma associated with them. In fact, a widely used drug in nightmares associated with PTSD, prazosin, acts mainly as a blocker of noradrenaline in the brain. Although a recent study suggests prazosin is no more effective than a placebo, I and my colleagues have seen several patients successfully treated with this drug.

So, certainly in some people, it is the dream content itself that has significance. It is not simply the random noise generated by our brains in REM sleep. And, indeed, dream analysis experiments confirm our own individual experiences – that dreams, while rarely rerunning the events of the day before, often have similar emotional themes, such as worries, anxieties or anger, as those experienced the previous day. Our daytime world colours our nocturnal lives. But is the precise content of our dreams important? Well, there is perhaps a little evidence to support this. Subjects who are given the task of navigating a maze in virtual reality performed much better if they reported dreams of moving through a complex environment than those who had other types of dreams, although crucially these were not the dreams of REM sleep.

Perhaps the hypothesis that attracts me most is the theory of REM sleep and dreaming proposed by Hobson and his collaborator Karl Friston, a serious heavyweight of the neuroscience world. I am not sure why this theory entices me so much. It may be that it shares parallels with the first paper that I read on sleep as an ignorant but interested medical student, described in the introduction to this book, by Crick and Mitchison. They argued that REM sleep was the process of reverse learning, cleansing the neural networks of unnecessary connections formed through our experiences during the day, and that dreaming represented the rubbish being taken out. Hobson and Friston's hypothesis is somewhat different, and another of its attractions is that it really unifies the worlds of REM sleep and dreaming, and in some respects the body and mind. It explains a purpose to our dreams. The theory is somewhat complex, and despite reading the papers repeatedly I am not entirely convinced afterwards that I fully understand it. This is the real reason I call Hobson on the telephone; I am seeking an idiot's guide to their theory. When I tell him I am not sure I totally understand it, he jokes: 'Neither do I!' But he does go on to provide an explanation.

Essentially, how we understand the world around us is based upon the interpretation of vision, feeling, sound, movement and experience. In order to do this, our brains, complex machines that take these inputs and ascribe meaning to them, need to have a model of the world around us. To some extent, that model, that basic circuitry, is genetically endowed. We are born with it. Indeed, it is present before we are born. Hobson tells me, 'The brain doesn't simply respond to external stimuli; the brain has a strong set of expectations. And you can say those are learned, but what Friston and I are saying is that they are learned before learning, they are genetically programmed.' But that model needs to be tweaked, constantly adjusted, and develops through life to define who we are as individuals. It determines whether we like wine, if we prefer Dalí or Constable, or how we respond to our partners when we have an argument. Hobson continues: 'The job of the brain is to anticipate reality, and it can't do that passively; it does that actively. It has a set of assumptions which then adjusts according to data. You and I have different ways of thinking because we are different people. And we have had different experiences. Our brains are probably not all that different, but the models that we have of the world are quite different.'

Essentially, this model defines our consciousness, but the tweaking can only happen when we are offline, detached from the external world, unable to move, rendered asunder from the world around us, even cut off from the processes through which we regulate our temperature. During REM sleep, Hobson and Friston believe that the brain integrates our experiences into this model, refining and moulding it, and our dreams represent a virtual reality environment in which this takes place. Ultimately, our dreams represent an amalgam of our experiences, our own cumulative model of the world rehearsed in a million different ways, all in the pursuit of understanding our own personal world and determining our individual consciousness. 'Your dreams are the subjective experience of the running of that model and that's what ties our work to psychoanalysis. The interpretation of dreams could be thought of as a way of understanding the model that the person has of the world. And the model that the person has of the world is clearly a function – in part – of his experience,' Hobson tells me.

In a way, this theory is also inclusive of Freud's view that our dreams are linked to childhood experiences, although Hobson implies Freud's mistake was to make it all about sex. 'I think what the Freudians have done is to sexualise something that is only part sexual and a very large part something else.' He has been keeping a dream diary for decades now. 'My sexual dreams are about 5 per cent of my dreams, by the way. If you think that all the other 95 per cent are governed by my Oedipal wishes, that seems to me absurd.'

I think that Christian's dreams of his youth represent something about his views of his childhood. He later tells me that his friends comment that there is an air of sadness about him, a tendency to hark back to younger, happier days, a loss of that contentment in earlier years. Rather than illustrating unvoiced sexual desires, both he and I believe that these dreams represent a wish to be back in those simpler years, unaffected by the challenges of adulthood and his narcolepsy.

And so it makes sense that some of our dreams may reflect the experiences of the day before, while others may bear no such meaningful relationship. They are the summation of our life to date. It is also unsurprising, therefore, that some dreams or nightmares may recur, in an effort to understand the meaning of our daytime experiences. But these dreams are a function of the brain. 'The brain creates the mind and the mind creates the brain,' Hobson says. The brain and the mind are one and the same. Cartesian dualism is dead.

* * *

Christian also describes prominent lucid dreaming, like Evelyn in Chapter 9. He has some degree of consciousness or awareness of his dreams, and can to some extent influence them. I think he rather enjoys it. He feels he has established control over his hypnagogic hallucinations, the succubus that previously visited him every night – 'over time I have battled it, and made it go away' – but now he can also direct his night-time life too. 'I can really easily lucid-dream,' he tells me. 'I can give you a really good example. I've never been to Hong Kong, I'm only going by films or video games that I've played or seen. In the dream I was in my version of Hong Kong – narrow backstreets, with all the big adverts for Coca-Cola – and I remember walking around those streets. I was staying at a hotel and I could literally walk around the hotel and go to the bar. I remember I went to the bar of my own accord.' I ask him if he genuinely felt that he was in control. 'There are many aspects of that dream that I wasn't in control of, because the nature of dreams is quite fantastical. But yes, I was able to walk around, communicate with people. I went to the bar, then walked out of the hotel. I remember going into a shop to look at clothing on a clothes rail. And then just kind of picking through this clothing. But I was actually in the dream, controlling it like it was reality.'

For some, lucid dreaming is a curiosity, or even a spiritual experience. It has frequently been dismissed as a figment of the imagination. It does have clear neurobiological markers, however. Lucid dreaming has been reported as arising after strokes in the thalamus, deep in the centre of the brain, but it can also be detected objectively, not just through people describing it. Monitoring the brainwaves of lucid dreamers using EEG as they move from non-lucid to lucid REM sleep shows changes in the frontal regions. Moreover, in a quite remarkable study, researchers have been able to definitively prove lucid dreaming. Taking six regular lucid dreamers into a scanner, the researchers asked the subjects to signal the onset of lucid dreaming. Remember that we are paralysed in REM sleep, with the only muscles not affected being the muscles that move the eyes, and those that allow us to breathe. As the subjects entered into lucid sleep, they signalled this in the only way they could, by moving their eyes in a pre-agreed combination. Left-right-left-right. The subjects used this signal not only to indicate the onset of lucid dreaming, but were asked to dream of clenching one hand for ten seconds, signal again before dreaming of clenching the other hand, and to continue doing this for as long as they could. Clearly, they could not really clench their hands, which were paralysed like the rest of them. Two of the subjects managed to complete the task. And in one, the results were staggering. The subject who lucid-dreamt of clenching his fist showed increased activity in the sensorimotor cortex on one side, which switched over to the other side when he also dreamt of switching sides. The activity in the sensorimotor cortex was similar to that seen when performing the task in wake, with real movements of the hand, thus clearly demonstrating that lucid dreaming is a very real phenomenon, with features of wakefulness while clearly in dreaming sleep.

Like many sleep phenomena, lucid dreaming represents another of these dual brain states. It appears to be an overlap between wakefulness and REM sleep, in the same way that sleepwalking does between wakefulness and deep sleep. When put in these terms, it is perhaps not surprising that people with narcolepsy report lucid dreaming much more frequently than those without the condition. If Christian often spends time hovering between wake and REM, which is what the hypnagogic hallucinations, sleep paralysis and vivid dreaming represent, then the fact that he lucid-dreams is almost to be expected. In fact, up to 80 per cent of people with narcolepsy report lucid dreaming.

The real significance of lucid dreaming, however, is that it provides a playground for dream researchers to understand the significance of dreams. If you can consciously influence your dreams, then this provides a natural experiment, to see how the content of your dreams relates to your daytime life. Consider a new skill, like learning to play the piano. As a lucid dreamer, if you push your dreams towards playing the piano, perhaps this might mean you learn it more quickly. Or the artist lucid dreamer, whose creativity and talents are boosted by lucid dreams of painting. The possibilities are endless.

* * *

Treating Christian's narcolepsy has been problematic. For him, the biggest issue is not the cataplexy but the excessive daytime sleepiness. Even small doses of stimulants have given him side effects. He hates the sensation of taking these medications, and says they make him feel 'chemical'. 'I've taken recreational drugs in my lifetime, so I understand the sort of feeling. And I get that from it [the stimulants].' He feels a little high, but the doses that we have been able to give him without accentuating this sensation are insufficient to properly keep him awake. He often feels 'wired', but not enough to stave off sleep. So, while he will occasionally use these medicines, he is reluctant to take them on a regular basis. And he is equally reluctant to use stronger, different types of medications, like the sodium oxybate that has made such a difference to Phil. 'Well, at the moment I'm able to manage my life. I'm lucky in this country, that I cannot work but still obviously receive some benefits. And in some other countries, you wouldn't even have any sort of treatment. But during the days, I can go to sleep – and generally I do two or three times a day.'

The daytime naps are a common way to treat narcolepsy, even alongside treatment with medications. Many of my patients with narcolepsy have planned naps – predetermined times at school or at work when they retire to a quiet room, or sometimes even a storage cupboard or toilet cubicle, and nap for ten to twenty minutes. One of the hallmarks of narcolepsy is the refreshing nature of these brief sleeps, recharging the batteries and allowing people to feel awake for the next few hours. He continues, 'I'm able to manage my life, housework, shopping, chores that people consider to be normal living – I'm able to manage that because I haven't got the pressures of a job.

So, basically, I receive my current prescription, but I'm not filling myself up with it every day. I take it in the mornings if there's something important going on, something that I really need to be conscious for.'

Christian also voices a sentiment that I occasionally do hear from some of my other patients with narcolepsy: there are some aspects of the condition that he rather likes. 'If you've seen the film *Inception* with Leonardo DiCaprio, that will give you a very good example of what my dreams are like.' On another occasion he tells me, using another film metaphor: 'It's like the Matrix exists, well within the confines of my own mind. Of a night-time I will just live another life.' I ask him if, in his dreams of the Matrix, he is Keanu Reeves or Laurence Fishburne. He chuckles: 'No, I'm just me. I'm the hero of the story that's being told. And I do kind of enjoy that.'

For Christian, I suspect the enjoyment he gets from leading multiple lives at night to some extent makes up for the limitations that his narcolepsy places upon him in his waking life. I tell him that I have one other patient who tells me that he feels that the dreams caused by his narcolepsy have some spiritual significance, that he communicates with the world on a different plane of existence. 'Yeah, I do feel kind of special, if that makes sense. And because I've got such a suspicion of certain areas of science and government, I do read stuff. I don't necessarily believe it, but I do take concepts on board of telepathy and stuff like that. And in my mind, if it is a real thing, I am going to be quite a good candidate to be able to do that. I do feel that I'm trying to communicate with people in my sleep, you know?'

I am not sure how much Christian really believes of this, but he clearly spends a lot of time trying to interpret his experiences, and make sense of them. What *is* clear is that there are some benefits to this personal nocturnal world. 'I see things that no one has ever seen, and no one's ever likely to see. Even though they're not real, I've still experienced them and seen them. The average Joe is never going to experience that because the average Joe doesn't remember. They might remember one or two dreams in a lifetime, you know?'

As for why we dream, when I look back through this chapter, I see lots of questions, but few definitive answers. REM sleep and dreaming probably have multiple purposes, and these are probably different at various stages of life. But, for the moment, the answer is still, 'I don't know.'

14

LOSING SLEEP

In 2005, the *New Yorker* magazine published an exposé on the regime of enemy combatants imprisoned in Guantanamo Bay (or 'Gitmo'). It detailed the techniques used by medical and scientific personnel 'intended to exploit the physical and mental vulnerability of detainees'. The practices to which the prisoners were exposed were based upon a Pentagon-funded programme called SERE – 'Survival, Evasion, Resistance and Escape' – developed originally by the US Air Force to help pilots shot down in the Korean War to cope with extreme abuse if captured.

At the core of these techniques applied in Gitmo was sleep deprivation and disruption of sleep patterns. The author of the piece writes, 'Sleep deprivation was such a common technique . . . that the interrogators called the process of moving detainees every hour or two from one cell to another "the frequent flier program".' The aim of this practice was to psychologically stress detainees to such an extent that they would lose the ability to 'self-regulate', or, as the article described it, 'the ability to moderate or control his own behaviour'. This loss of self-control would facilitate interrogation, softening the prisoners up and making them more likely to reveal important morsels of intelligence.

The argument as to whether systematic sleep deprivation in this context represents a legal and ethically valid form of interrogation or frank torture rages. It leaves no wounds or scars, and causes no pain. The United Nations, in its Convention Against Torture, defines torture as

> any act by which severe pain or suffering, whether physical or mental . . . intentionally inflicted on a person for such purposes as obtaining from him . . . information or a confession . . . when such pain or suffering is inflicted by or at the instigation of or with the consent or acquiescence of a public official or other person acting in an official capacity. It does not include pain or suffering arising only from, inherent in, or incidental to, lawful sanctions.

It is this last phrase, 'lawful sanctions', that is vague, open to interpretation, and debated. Sleep deprivation would certainly qualify as an act to cause suffering.

Regardless of how sleep deprivation is currently defined, it has been utilised in the arsenal of interrogation or torture techniques for centuries. First properly documented in the late fifteenth century as being used in the Catholic Inquisition, it has been relied upon through the ages, from the witch-hunters of sixteenth-century Scotland to the interrogation camps of the KGB, and no doubt still in various dark corners of the world to this day.

And although it is true that sleep deprivation leaves no physical traces, not only can it leave psychological scars and mental pain, it is also potentially highly dangerous. While in humans systematic sleep deprivation for long periods has never been properly scientifically explored, in animals it has proven fatal. Dogs kept awake will invariably die after 4–17 days. Similarly, rats die after 11–32 days awake.

Imagine for one moment that you were subject to this torment. You crave nothing more than a few snatched minutes of sleep; you cannot think clearly, your vision is blurred, your limbs ache with fatigue. It is not a Catholic Inquisitor or a Guantanamo guard shaking you awake at the first sign of dropping off, however. It is yourself, your own brain: you are your own torturer. And that is insomnia.

* * *

When Claire first walks into my consulting room, she is in trouble. She does not look it, though. At first sight, she is immaculately dressed, in her early fifties, slim, and really rather beautiful. She appears just like many successful, wealthy women striding through the streets around London Bridge, a stone's throw from the City of London. However, for the past five years, she has been plagued by debilitating insomnia.

Her sleep initially worsened as she approached the menopause, but the precipitating cause of her insomnia in her own mind is clear. 'I was going back to work after fifteen years at home with the children,' she tells me. 'It was partly me being fifty and wanting to prove myself so badly in the workplace. I actually had a really low-paid job, although it was very high responsibility. I was on a mission to prove myself and show that I was worth it and of value.'

It sounds as though she put herself under enormous stress in an effort to impress, which was compounded by her battling for a pay rise. 'In the lead-up to the menopause I was struggling with sleep, but I would just say [to myself], "I have trouble sleeping. I wake up in the night. I'm just not a great sleeper." But I could still function during the day.'

The situation at work seemed to have tipped her over the edge, however. 'I stopped sleeping. I know that it sounds mad, but I really think for a year or so I got so little proper deep sleep that I stopped functioning.' I ask her more about her sleep patterns at that time. 'I would go to bed. But literally as I was walking upstairs I would be starting to get panicky. I just knew what was coming. And then my heart would start beating fast. Not a full-blown panic attack, but I could feel the adrenaline going through my body.'

The anxiety about the process of getting into bed would become a self-fulfilling prophecy; the fear of how difficult it would be to go to sleep added fuel to the fire of insomnia. Claire continues: 'So I'd lie there for a couple of hours and then I just knew that it wasn't going to happen. So I'd get up and go downstairs and make myself a cup of herbal tea, walk round the kitchen, keeping the lights quite low. And then I'd go back and try again.' But the distress of sleep always being out of reach took its toll, both physically and emotionally.

'I'd start to get quite upset. Sometimes I'd wake up my husband, crying and semi-hysterical – I hate to admit that. He would be very sweet and he'd try to calm me down. He used to say all the right things, he really did. And then towards the early hours I'd probably get some kind of very light, dreamlike sleep, but I'd wake up feeling wrecked.'

And so began the downward spiral into oblivion. Her lack of sleep made it even more of a struggle to perform to her own expectations at work, raising her anxiety levels further and making sleep even more elusive. 'And then I had a breakdown,' she says.

* * *

There is nothing quite like the loneliness of the insomniac, awake in the middle of the night while the rest of the world sleeps. Claire has kept a sleep diary, in which she writes:

'Everyone else in the house is asleep and I'm just despairing really because I've tried so many things, and every night I just find myself back down here [in the living room]. It just feels really lonely, like this is never going to end.'

But she is not alone. In fact, far from it, for insomnia is unbelievably common. If, like Claire, you have found yourself unable to drift off to sleep, with difficulty staying asleep, or waking up feeling that you have slept very poorly, you are in a very large club. Insomnia is by a significant margin the commonest condition affecting sleep. Roughly a third of adults will report features of poor sleep, and about one in ten adults will have chronic insomnia, resulting in an ongoing poor experience of sleep, coupled with daytime consequences such as fatigue, irritability, difficulty concentrating and lack of motivation. But insomnia is not just a medical condition, it is also a symptom – of medical problems like an overactive thyroid gland, or as a result of medications. Insomnia can be a feature of psychiatric disorders like anxiety, depression or bipolar disorder. In fact, 50 per cent of people with insomnia have a psychiatric diagnosis (although that also means 50 per cent do not). Even insomnia as a medical disorder, in the absence of any other underlying problem, is an umbrella term. There are different types of insomnia, and although it sounds strange, not everyone with insomnia is deprived of sleep.

For some people, their experience of poor sleep is not borne out by the evidence. It is rare that I bring patients with insomnia into the sleep laboratory. Someone who does not sleep very well at home will definitely struggle to sleep covered in electrodes, in a strange bed, knowing that their every move is being recorded and analysed. Occasionally, however, when there are questions as to the cause of their insomnia, or I am wondering about another type of sleep disorder, I will admit patients for a night.

It is incredibly common, when we meet after the sleep study, and I ask how they slept, to hear: 'I slept terribly that night.' But when we look at the sleep study, it shows a very decent night's sleep – seven or more hours, with lots of deep sleep – despite the person in front of me being adamant that they only slept for an hour or two. This type of insomnia is termed 'sleep state misperception' or 'paradoxical insomnia', and is likely to explain why there is a huge overlap between the sleep studies of patients with insomnia and normal sleepers. Something about the way the person experiences sleep is different. Perhaps it is the quality of sleep, something we cannot gauge with our standard technique of measuring sleep, the polysomnogram. Or maybe it is as simple as the brain filling in time between the brief awakenings that are a feature of normal sleep, perceived as wakefulness by someone predisposed to this form of insomnia, rather than the deep sleep seen on the sleep study.

For others with insomnia, their sleep may be broken, disrupted several times per night, but their total sleep is a normal amount. Even those with a reduced total sleep duration may actually have normal amounts of deep sleep, that stage of sleep most important to physical restoration and refreshment.

For the hardcore insomniac, like Claire, however, there is clear evidence of very curtailed sleep, sometimes only a few snatched hours every night. And it is in those people with a short sleep duration that we see clear biological markers of stress, what is termed 'hyperarousal'. That jangling of the nerves, the racing heart, being on full alert, the feeling of being excited or vigilant – all of these are features of hyperarousal. When under stress, a number of neurotransmitters and hormones come into play. The state of being stressed or anxious leads to the boosting of a number of systems in the body, resulting in higher levels of cortisol, adrenaline and noradrenaline.

When these systems are studied in patients with insomnia, crucially with a short sleep duration, we see increased levels of the breakdown products of these hormones in the urine. This category of insomniacs also demonstrates other features of this 'hyperarousal state' – increased nocturnal heart rate, increased oxygen consumption (implying a higher metabolic rate) and increased pupil size, once again a measure of the heightened activity of the sympathetic nervous system, which mediates the 'fright-fight-flight' response. Importantly, these changes are not seen in people with insomnia who are actually getting a reasonable amount of sleep.

Insomnia and sleep deprivation are often conflated. The health risks of sleep deprivation, those people who simply do not allow themselves to sleep enough, are well documented – mortality, weight gain, high blood pressure, diabetes . . . the list goes on and on. And so it is natural for people with insomnia to worry about these issues. Surely decades of poor sleep would give rise to the same damaging consequences on health? However, sleep deprivation and insomnia are quite different. If you study someone who is sleep-deprived in the sleep lab, they doze off very quickly, and while they are awake they perform poorly in tests of vigilance. In stark contrast, insomniacs with short sleep will take much longer to drop off to sleep, and are much more vigilant while awake.

But here, the distinction between the different types of insomnia – with short sleep duration and normal sleep duration – is of fundamental importance. Both types of insomnia are associated with the brain being more active than normal. Using imaging techniques and monitoring brainwaves, both groups of insomniacs are found to have increased brain activity in sleep, which may explain why patients with sleep state misperception or a normal-ish sleep duration experience sleep as wakefulness or have sleep that is not refreshing. But it is only those with a short sleep duration who have this heightened activity in the whole body, as demonstrated by chemical and physiological markers like heart rate. While the effects on brain activity are important in terms of the subjective experience of sleep, many of the health problems related to insomnia seem limited to those with a short sleep duration and the physiological hyperarousal affecting the whole body, rather than just the brain. Studies of cognitive performance in people who say they have insomnia do not show major differences when compared to normal sleepers, but when you separate those with normal amounts of sleep, even poor-quality and broken sleep, from those with objectively measured short sleep, it is the insomniacs with short sleep duration who have significant cognitive problems. In contrast, people without insomnia who are sleep-deprived do not exhibit features of activation of these hormones, neurotransmitters and cardiovascular markers of hyperarousal, nor do they show cognitive problems to the same extent.

Similarly, when the risk of conditions like high blood pressure and diabetes are analysed in people with insomnia, those with properly measured short sleep have higher rates of these conditions, while those sleeping six hours or more have no increased risk at all. Insomniacs with short sleep have even been shown to have slightly increased mortality. But how this might occur seems to be different from people who are simply sleep-deprived. We know that sleep deprivation is associated with weight gain, so could it simply be that insomniacs sleeping very few hours gain weight, which then predisposes them to diabetes, high blood pressure and all the associated ills? Well, actually, chronic insomniacs with short sleep duration do not seem to gain weight any more than normal sleepers. In fact, if anything, these patients are less likely to become obese than normal individuals. Instead, it is likely that the chemical and physiological effects of hyperarousal directly contribute.

Cortisol, a natural steroid, is associated with increases in blood pressure and diabetes. We see this all the time in patients on steroids to suppress their immune system if they have an autoimmune neurological disorder. The increased activity of the sympathetic nervous system and related chemicals like adrenaline directly influence the heart and blood vessels, leading to the loss of the usual drop in blood pressure expected at night.

To summarise, the physical effects of insomnia with very curtailed sleep seem to be a marker for these physical changes in one's hormone and cardiovascular state, and look to be responsible for some of the risks to health that insomnia entails. For people with insomnia with a reasonable sleep duration, there is evidence of abnormal brain activation, but from a physical perspective, this group of insomniacs shares more with people with normal sleep.

So what drives this hyperarousal state seen in people with severe insomnia, those who sleep very little? Is hyperarousal caused by the short sleep of these insomniacs, or is the insomnia generated by this hyperarousal state? The answer is not entirely clear, but the lower levels of activation of these chemical and neurological systems in those with sleep state misperception – those people who have subjective poor sleep but have a normal total sleep duration – and those people without insomnia but who are sleep-deprived would certainly suggest that it is the hyperarousal state itself that causes these more severe forms of insomnia.

There are undoubtedly genetic factors at play. Insomnia often runs in families, and studies of twins have suggested that 57 per cent of insomnia can be explained by genes. A recent study has identified seven genes that contribute to insomnia, so a genetic predisposition to this hyperarousal state is certainly possible. It is quite normal for people under stress, due to a new job, a relationship issue, a death in the family, to have a period of insomnia, and a period of hyperarousal. But it may be that if you have the right genes, in the context of such a stressor, you are at higher risk of that state of increased excitability, mentally and biologically, persisting beyond the presence of that stressor. And that state of increased vigilance or arousal drives the insomnia to become chronic.

There are clearly also psychological factors in the mix. As I've already said, roughly half of patients with chronic insomnia have underlying psychiatric disorders, especially anxiety – and hyperarousal is a strong feature of anxiety. So the anxiety itself may be driving the insomnia. But what about those *without* anxiety, the other 50 per cent who do not have any psychiatric disorder?

For many of the people I see in clinic, they do not have panic attacks, are not worriers during the day. But many tell a common tale. They report feeling fine during the day, but at the point of going up to bed they start to worry, specifically about the process of going to sleep. They become anxious that they will not drift off to sleep, that they will struggle throughout the night. They live in dread of the night ahead. When their head hits the pillow, rather than associating their comfortable bed with the delights of drifting into blissful sleep, they see their bedroom as a place of torment, an instrument of torture. 'I feel exhausted as I get into bed, but as soon as the lights go off, my mind races and I feel wide awake,' is the common refrain. It is at that moment that they become hyperaroused, their brain feels 'wired', and sleep suddenly becomes out of reach. Like Sisyphus rolling his stone almost to the top of the hill before it slips out of his hands and rolls all the way down again, they are tantalisingly close to sleep when it is suddenly snatched away from them. The longer this persists, the worse their relationship with sleep becomes.

There is, however, another aspect to this state of heightened psychological and physical stress response. It makes people feel awful. Alongside the poor sleep, sufferers of insomnia can feel like they are on the verge of dying, and unless it is immediately obvious to you that there is a significant element of anxiety, it can contribute even more to the insomnia, because then you begin to worry that you have something seriously wrong with you. Claire says, 'I used to go online and google insomnia problems. I convinced myself I had certain syndromes.' I ask her if she was worried about any syndrome in particular, fully knowing the answer, having heard it many times: 'Fatal familial insomnia. I convinced myself I was going to die within six months.'

Fatal familial insomnia is a prion disorder related to Creutzfeldt–Jakob disease, or 'mad cow disease'. Caused by a faulty gene, passed down through generations, it is a relentlessly progressive neurological disorder, invariably fatal after an average of eighteen months. Patients present with insomnia that gradually worsens. As the disease progresses, there are changes to the autonomic nervous system, an inability to maintain blood pressure with marked fluctuations, high and low, sweating, and constipation. Delirium ensues with hallucinations and behavioural change, and in the latter stages sufferers drift constantly in a limbo between wake and sleep. This condition is, however, unbelievably rare, affecting only forty known families in the entire world. Perhaps it is indicative of the level of anxiety with which people with insomnia can be afflicted that they occasionally quickly jump to this diagnosis, convincing themselves of the other symptoms of this disease.

* * *

When I first meet Claire, she recognises certain personality traits in herself that might have led her to this point. Moreover, when we discuss her sleep prior to her current problems, she tells me that her sleep would always be somewhat affected by other things going on in her life. The trigger for her hyperarousal state is evident to both her and me: 'I just think the stress of work, the stress I put on myself for work, was self-inflicted,' she tells me. 'Because of this need to prove myself as an older woman, going back into the workplace. Somebody else might have reacted very differently to what was going on. But it's just in my nature to want to do a good job and I'm quite sensitive to any kind of criticism. So I would come home and agonise over them [criticisms] and replay them in my mind.'

She also recognises this hyperarousal state in herself. 'I just basically got to the point where I was not able to fall asleep. That switch that sends people to sleep stopped working for me. So whether it's a nap during the day or going to bed at night, it didn't work.' She describes that very typical difficulty in sleeping during the day and the increased vigilance of the hyperarousal state. 'My brain was trying so hard to function that it had to produce so much adrenaline that it was just permanently on alert. It wouldn't go to that place of rest.'

After a couple of years of terrible sleep, however, Claire hit a wall.

> That was when I asked for my pay rise. There were some hard negotiations which I just found too much. I spiralled downwards basically. And with the lack of sleep, combined with my reaction to those events, I went into a place of deep depression really, and I didn't know how to get

> out of that. And then the depression led to more insomnia and the insomnia made the depression worse. And my hair started falling out by the handful. I lost about half of my hair. But I don't think I recognised it for what it was for quite a long time.

By the time I meet Claire, she has already been diagnosed with anxiety and depression. She has been seeing someone for her psychiatric issues and, after cycling through various drugs, has finally been commenced on an antidepressant that has helped her a lot. 'I think about the fourth drug we tried suddenly worked really well. Within a few weeks I felt so much better. The sleep didn't automatically get better, it's just my mood changed so much. And it was like this cloud lifted. It made me realise that I think I've had depression to some degree since my children were born twenty years ago. Because it [the medication] changed everything.'

Despite the medication, there are ongoing issues with her anxiety and depression – and her sleep is still terrible. She has been trialled on several medications specifically for sleep, and some of these worked briefly before wearing off. Others have triggered a worsening of what was previously mild restless legs syndrome, an unfortunate complicating factor in Claire's case. She has also tried a psychological therapy called acceptance and commitment therapy, which focuses on teaching people to accept or embrace their insomnia, and thus aims to reduce the feelings of stress associated with their lack of sleep, but for Claire it has done nothing.

Her insomnia persists a year after her breakdown. At the moment, she gets into bed at 10 p.m., feeling exhausted, but at the point of getting into bed, she describes that familiar alerting response, that loss of the switch into sleep. Like a carrot on a stick, when she approaches the point of sleep, it constantly moves away from her. She lies in bed for three or four hours, seeking sleep, before finally giving up and going downstairs. By 3 or 4 a.m., the sheer exhaustion of having been awake for twenty-odd hours eventually overrides the adrenaline and cortisol pumping through her veins and brain. She finally drifts off to sleep, only to wake up without an alarm between six and seven in the morning.

* * *

The relationship between insomnia and psychological or psychiatric problems is complex. As I have already said, anxiety in some shape or form may underlie the hyperarousal state that is common to more severe forms of insomnia. Insomniacs with short sleep often have a particular psychological profile, with low mood, fatigue, and anxieties about health. Those people with sleep state misperception have a degree of overlap, with low mood and anxiety, but have a tendency to spend time ruminating, thinking about things over and over, as well as intrusive thoughts. These subtle differences in psychological picture, and the associated physical derangement of hormones and cardiovascular measures, have led some researchers to propose that these two types of insomnia are fundamentally different. Those insomniacs with normal sleep duration, i.e. sleep state misperception, do not have the bodily features of hyperarousal, and therefore few long-term consequences, and seem to be more responsive to treatment. Those short sleepers with physical hyperarousal – the amplified chemical and biological systems that mediate the 'fright-fight-flight' response – have a different psychological profile, are at increased risk of health problems related to their insomnia, and are more likely to be difficult to treat.

So, psychiatric problems can certainly give rise to insomnia. Ninety per cent of people with clinical depression have insomnia. Even as a medical student I was taught that early-morning waking is a hallmark of depression, but other types of insomnia – such as difficulty getting off to sleep and staying asleep – are also common. Patients with schizophrenia will often have terrible insomnia, and frequently the slipping into a psychotic episode will be heralded by increasing difficulty with sleep. But the relationship between sleep and psychiatric disorders also goes in the opposite direction. Insomnia itself is a risk factor in developing psychiatric disorders or can make these disorders more difficult to treat. The presence of insomnia, entirely independent of all other factors, significantly increases the risk of developing depression subsequently, and this is especially the case for those people with insomnia and a short sleep duration. And for those people with depression already, insomnia is a marker for increased suicidal thinking and increased risk of relapse back into depression. Patients with depression are more refractory to treatment when insomnia complicates matters.

There are a number of unanswered questions, however. This is a scientific field in its infancy, and we do not yet fully understand either the intricacies of the relationship between sleep and mental health or what underlies it. Both insomnia and psychiatric disorders result in changes to both brain circuitry and brain biochemistry, and so it is not surprising that changes in sleep or mental health may have knock-on consequences on one another. They are both the chicken and the egg. But it may also be that there are shared genetic factors that predispose both to insomnia and psychiatric issues, to add an additional layer of complexity to this area of research. Whatever the true nature of this relationship, it highlights the need for psychiatrists and sleep physicians to think holistically, to not simply focus on the problem that they are most familiar with, to approach these patients without the blinkers of having been trained in one particular specialty.

* * *

Strategies for the treatment of insomnia have historically focused on medication. Benzodiazepines hit the market in the early 1960s, and rapidly became the staple of treatment for insomnia and anxiety, popped in vast amounts. So mainstream were these drugs that the Rolling Stones, not known for their reluctance to indulge in recreational chemicals, wrote an ode to Valium entitled 'Mother's Little Helper'. Over the past few decades, however, the dangers of benzodiazepines and related drugs like zolpidem and zopiclone have become apparent: risks of morning sedation, road traffic accidents, falls and hip fractures; the triggering of sleepwalking and other non-REM parasomnias; and, importantly, withdrawal effects and dependence, with ever-increasing doses required to get the same effect on sleep.

As a result, there has been a shift away from these types of drugs to other agents such as melatonin, antihistamines and sedating antidepressants. They all have their problems, different side effects and a wearing-off effect, but used in the right way they can make a big difference.

Most alarmingly, there is growing evidence that points to hypnotic drugs, particularly benzodiazepines and related drugs, increasing the risk of subsequent dementia. However, as with many questions in sleep, this is a complex area. We have already discussed the glymphatic system, a network of channels in the brain responsible for clearing waste products, akin to the lymphatic system in the rest of the body. In deep sleep, these channels open up by up to 60 per cent, allowing the carriage of potentially toxic substances like beta-amyloid, the protein implicated in the process that underlies Alzheimer's, away from the brain. Deep sleep facilitates housekeeping of the brain, and if you are sleep-deprived, due to insomnia or otherwise, this process of brain cleansing is affected. Sleep deprivation results in increased levels of beta-amyloid in the fluid within these channels, implying reduced clearance of this protein and other toxins into the cerebrospinal fluid, the liquid that bathes the brain and spinal cord. So sleep deprivation, and insomnia with a short sleep duration, could well in itself carry an increased risk for Alzheimer's disease.

There is yet another possible explanation, though. Many degenerative disorders of the brain result in subtle changes years or even decades before symptoms become more obvious. Consider John and his dream-enactment behaviour, often a precursor to Parkinson's disease by many years, and the association of anxiety with this condition. It may be that, in Alzheimer's disease, early changes to the biochemical pathways in the brain years before memory deteriorates result in worsening sleep or anxiety. It may be that the insomnia is not the cause of Alzheimer's, but the result of the very early stages of the disease.

The reality of this association between poor sleep and Alzheimer's disease is yet to be fully unravelled, but whatever its nature, concerns regarding this and the other side effects of drug-based treatments has led to a sea change in how we treat insomnia. Specifically, there has been a massive shift towards non-drug-based treatments. The most widely studied and used of these is a treatment called cognitive behavioural therapy for insomnia, or CBTi. This treatment uses behavioural techniques to essentially reprogramme the brains of people with insomnia.

In the 1890s, Ivan Pavlov, a Russian scientist, began experimenting on dogs. He had noticed that his dogs began salivating whenever he came into the room, in expectation of food. He noted that while salivation in response to food is a hardwired response, the association of him with food must have been learned. He then began to train his dogs to associate a bell being rung with the presentation of food, and rapidly his dogs began to salivate at the mere sound of the bell. This phenomenon is termed conditioning, a learned response, and humans are in some ways no different from Pavlov's dogs, in that we are also subject to conditioning. For example, think of a drink or food that in the past has made you ill. It is likely that even the sight of that food or drink now makes you feel nauseated, perhaps even the mention of it. Conditioning also applies to sleep, however.

If you have no problems with sleep, you associate your bedroom with feeling sleepy, that comforting sensation of slipping under the duvet, putting your head on the pillow, and that wave of relaxation as you drift off to sleep. For the insomniac, the conditioned response is very different. The bedroom is a location of stress and anxiety, dread of the night ahead. The bed itself is a cause of mental or physical hyperarousal, making sleep even more difficult. As Claire says: 'It [my bed] is a place of misery and a chamber of torture now.'

At the core of CBTi is the breaking down of this negative conditioned response to one's bed, and rebuilding a positive association. The trick is to once again establish the bed as a sanctuary rather than a torture chamber. This is done by a number of ways. The first is by a rigid regime to avoid you lying in bed at night for prolonged periods while awake, forcing you to leave the bedroom after a few minutes of struggling to get to sleep, and to avoid using the bedroom for anything other than sleep. The second initially sounds counterintuitive. After all, the last thing an insomniac wants is to be sleep-deprived. But what many people with insomnia do is compensate for their poor sleep by spending more time in bed. And what that achieves is to increase the time they are in bed not asleep, thus strengthening the negative conditioned response. So, by limiting the time allowed in bed to five or so hours for a period of a couple of weeks, this builds the brain's drive to sleep. Like Claire hitting 3 or 4 a.m., eventually the sleep deprivation overrides the hyperarousal state and sleep ensues. This is a crucial step in rebuilding the association between bed and sleep.

At its most extreme, sleep deprivation as a treatment for insomnia has been developed in Australia into an experimental technique called intensive sleep retraining. The patient is asked to stay in bed no more than five hours the night before they come into the sleep laboratory. At 10.30 in the evening, the protocol commences. For the next twenty-four hours, every thirty minutes the patient is allowed to try to go to sleep, with electrodes attached to their scalp. After twenty minutes, if they do not fall asleep, they are asked to get up. But if they do fall asleep, as proven by their brainwaves, after three consecutive minutes of sleep they are woken. By the end of the 24-hour period, they have had forty-eight opportunities to fall asleep. In theory, by the end of the protocol, they are so sleep-deprived that they fall asleep as soon as they are allowed. The conditioned response between bed and sleep has been re-established. This technique sounds like something straight out of Guantanamo Bay, but results from trials have been impressive. This short sharp shock rapidly reconditions the response to getting into bed, and results in quick improvements in sleep.

CBTi uses similar principles, but in a less brutal manner, and is combined with other techniques to induce relaxation and address hyperarousal, as well as the teaching of good sleep hygiene, the term for behaviours surrounding sleep such as avoidance of bright light and caffeine, and adequate wind-down periods. The results of CBTi are equally good. The improvement in sleep in the short term is equivalent or even superior to hypnotic drugs, and the benefits have been shown to persist for up to three years.

As a result, CBTi is usually recommended as the first-line treatment for insomnia, sometimes in combination with medication in the short or medium term. CBTi can also be used to help the gradual withdrawal of long-term drugs for sleep.

When Claire and I discuss treatment options for her ongoing insomnia, we decide to go down this route of treatment, avoiding adding in further medications in the first instance. The response is quite remarkable. We meet a few weeks after starting CBTi. 'I was sceptical because I'd tried so many things,' she admits. 'I was very up for drug help. I just didn't believe that anyone could help me.' She has seen the sleep therapist twice now, and is following the CBTi programme to the letter. She is now spending no more than seven hours in bed.

> I count the minutes until eleven o'clock because it's so hard to stay up and I'm desperately sleepy. I'm waiting. I have my wind-down time. I spend an hour running a bath and lighting candles. And then in the morning the alarm goes at six and I have to get out of bed immediately. I go straight downstairs and I sit there feeling a bit sorry for myself. And I've done that every day religiously for three weeks. And it's got better and better and better. I almost can't believe that this is happening, but I'm sleeping. The switch [into sleep] seems to be re-emerging!

It is almost too good to be true, and both the sleep therapist and I are cautious not to be overoptimistic. But Claire feels transformed. 'I almost can't remember the last time I felt this human. Full of energy and able to concentrate. I feel so excited by what life now holds,' Claire continues. And, despite my reservations, Claire continues with the rigid CBTi programme for three months, during which her insomnia is held at bay. With improvements in sleep, her restless legs also ameliorate.

The next few months, however, are rocky. For no obvious reason, she suddenly has a resurgence of her anxiety, and her sleep deteriorates again. She ascribes this to a loss of discipline with her sleep programme, but I am not so convinced. I think this is part of her tendency to have incredibly high expectations of herself, to take responsibility for everything around her and her reluctance to accept that she does not have control over everything in her life. We up the dose of her medication for her depression and anxiety, and I suggest we try again to address her psychological difficulties through a clinical psychologist. And, slowly but surely, she gets better. With the clinical psychologist, she recognises patterns of behaviour and thinking that place her under undue stress, pressures that she creates for herself, negative thoughts about her life and her achievements. By recognising these destructive thought processes, she has learned to lessen the emotional consequences of minor events in her life that previously would have triggered profound stress. She has also found a complementary therapist who has helped.

When I talk to her some nine months after we first met, she tells me she is 'fabulous'. Her anxiety and mood issues have lessened, and she is sleeping regularly. When her head hits the pillow, she drifts off to sleep. Without medication. She is still on a low-dose antidepressant, but is even reducing this gradually. For the first time in five years, she feels normal.

* * *

Ask anyone with insomnia what sleep means to them, and the importance of sleep is immediately obvious. It affects every aspect of our waking lives – mood, energy levels, cognition, memory, immune system, metabolism, appetite, anxiety levels. It influences our relationship with everyone around us, and with ourselves. As Claire says: 'Sleep changes everything. Without it, no one can function. You can try for a bit, but it gets to the point where you actually start shutting down. I felt like my body was shutting down and my brain was shutting down. And suddenly, with sleep, it's all opening up again.'

EPILOGUE

SOME GENERAL THOUGHTS ON SLEEP

You can survive longer without food than without sleep. The fact that sleep is fundamental to life is unarguable, but in modern society, at least until recently, we have taken for granted that sleep simply happens, and is a necessary evil to allow us to live our waking lives. Recently, however, largely through the efforts of many of my colleagues around the world, there has been a shift in how we view sleep. Rather than being a hindrance to our working and social lives, a biological process that keeps us from being productive, the concept of the importance of sleep is percolating through. Its role in the maintenance of our physical and mental health, our sporting prowess, our cognitive abilities, even in our happiness, is slowly being appreciated. And rightly so. People are taking sleep seriously.

We have moved a long way from the *Egyptian Dream Book*, and indeed from Freud's *Interpretation of Dreams*. Freud of course was not even aware of the existence of REM sleep, discovered half a century later, but our understanding of sleep and its disorders, even in the past fifty years, has exploded. There is an apocryphal-sounding but true story of the inventor of CPAP – the technique used to treat sleep apnoea – being invited from Australia to Edinburgh in the 1970s to give a keynote lecture at a conference. At the end of his talk, the professor of medicine at

Edinburgh stood up and announced to the audience, composed of the great and the good of the UK medical establishment, that sleep apnoea did not exist in 'this country', and perhaps 'we have exported it all to the colonies'. The eminent Australian physician was so shocked and insulted that he did not return to the UK for some twenty years.

Nowadays, such a statement would be unthinkable, even by the most medically ignorant. And the column by the general practitioner published by the *British Medical Journal* in 2013, in which he stated that restless legs syndrome was a condition made up by pharmaceutical companies to flog drugs, would, I hope, not be published today. We have moved away from sleep and dreaming being a spiritual phenomenon to one firmly rooted in the physical realm, with an underlying neurological basis. We have fundamentally understood that sleep disorders derive from neurological, psychiatric and respiratory dysfunction, rather than God, witchcraft or lunacy. And we strongly grasp the necessity of sleep in the regulation of neurological, psychological and cardiovascular health.

This shift has driven, and also been driven by, the view that sleep medicine is a multidisciplinary field, requiring the involvement of neurologists, respiratory physicians, psychiatrists, cardiologists, psychologists, ear, nose and throat surgeons, and dentists. And that sleep research necessitates scientists specialised in all these fields.

Technological advances in studying sleep have also been crucial to this scientific progress. It started with the EEG in defining the different stages of sleep, but we now have ways of measuring airflow, chest movements, body movements. We have the ability to image the brain, not only its structure but also its function, using techniques like functional MRI, and radioisotope-based scans like PET and SPECT. We can analyse electrical activity deep within the brain, through implanted electrodes. We can monitor fluctuations in hormones, genes, proteins and metabolites over a 24-hour cycle. We have laboratory techniques that can switch genes off and on simply by shining a light on a section of brain. We can breed mice with manipulated genes, and we can study the genomes of huge numbers of people, looking at common genetic variants and their association with different facets of sleep and its disorders. We can relatively cheaply unravel the genetic code of an individual, identifying each distinct position in the 3-billion-long sequence of letters that comprise the human genome, finding rare mutations that cause disease. Many of these tools now at our disposal were totally unimaginable only a few years ago.

* * *

Two of the most frequently asked questions I hear in my sleep clinic are, 'How much sleep is enough?' and 'What do you think of my sleep tracker?' With regard to the first question, I do not answer it, at least not with a number of hours. Indeed, I *cannot* answer it in that way. The question is similar to 'What is the normal height for a ten-year-old?' If I look at my daughter's class photo, the children range in height hugely, but all of them are normal. Likewise, there is a range of normal sleep requirements. It depends on your genes, and the quality of your sleep. The right amount of sleep is the number of hours needed for you to wake up feeling refreshed, not sleepy during the day, but then ready for bed at a regular time, without difficulty dropping off. If you are achieving that regularly, if you are waking up before your alarm, and not catching up on sleep when you have the chance at the weekend, then you are getting the right number of hours' sleep.

As to the second question, I usually tread delicately for

fear of offending. We live in an age where everything needs to be measured. We constantly feel the necessity to apply metrics to our lives, be it how many steps we take, how many Instagram followers we have, how much we earn, how many calories we consume, and of course how much sleep we get. But I do wonder if this tracking of our sleep quantity is helpful. If you meet the criteria above, i.e. have sufficient sleep to function properly during the day without feeling sleepy, then the likelihood is you are getting enough sleep. If you feel tired and unrefreshed, you probably are not, and don't need a sleep tracker to tell you this. But, apart from just wasting some money on the gadget, there are other potential downsides. These devices are at present relatively inaccurate. Apply five different trackers to your arm and you will get five widely diverging estimates of your sleep duration. They measure movement, not sleep, and depending on the algorithm, give an inaccurate measure of sleep of varying degrees.

Where they could help is if they can prove that your insomnia is related to sleep state misperception, that your experience of insomnia is not borne out by limited sleep overnight. That, however, requires the data to be reliable, for you to trust that the device you are wearing gives a very accurate picture of your sleep. And there is a further issue. If you are already worried about your sleep because you have insomnia, then constantly tracking your sleep can exacerbate these worries – can intensify your obsession with your sleep, and thus make the problem worse. This phenomenon now has a term – 'orthosomnia' – where people are diagnosing themselves with sleep disorders based upon the dodgy output of their sleep trackers. For most people, sleep is a subjective experience, and insomnia is most often associated with a normal sleep duration, so your sleep tracker telling you that you have light sleep when you expect deep sleep can in itself have profound effects on your own perception of your sleep.

This does not mean that these devices are totally devoid of value. They can track improvements or deteriorations in sleep with interventions like cognitive behavioural therapy for insomnia, and perhaps most importantly can provide researchers with 'big data' on sleep patterns, where the noise of an imperfect technique to quantify sleep is diluted out as a result of huge numbers of individuals. But for the individual person, especially those sitting in front of me in my sleep clinic, I have my doubts. The obsession with the number of hours in bed, with inaccurate measures, results in us overlooking what I have shown in the pages of this book: that there are many factors – biological, psychological, behavioural, environmental – that all influence sleep quality as well as sleep quantity.

It is also important to stress that most studies looking at relationships between sleep and physical and mental health are imperfect. If there is one thing that is very obvious from hearing some of the stories I have told you, it is that we are often unreliable witnesses to our own sleep. Our perception of our night-time and the reality are often entirely different. One of the major problems with performing these studies is that for the most part we rely on what subjects within these studies tell us. Performing sleep studies on everyone is inordinately expensive and practically impossible, at least using currently available technologies. So when we look at large populations, and correlate sleep factors with high blood pressure, heart disease, Alzheimer's disease and so on, it is likely that a group labelled with insomnia or sleep deprivation is a mixed bag. If we take the example of insomnia, some will have a normal sleep duration, others will have a short sleep duration, some will be on medication causing their insomnia and yet more will have sleeplessness due to the pain or breathlessness caused by medical problems. When correlating sleep duration to mortality, although we try to adjust for known factors, it is almost entirely impossible to adjust for everything, or to be sure that everyone reporting a sleep time of over seven hours actually is sleeping that amount.

Please don't misunderstand me: I do not for a second doubt that sleep problems have really major consequences on our physical and mental health. I just believe that there are subtleties there that we have yet to fully appreciate, nuances that we have not yet uncovered. Perhaps as the technologies available to us improve, and we can track sleep rather than movement in the long term, these intricacies will become clearer.

There is another important message in this book. The days when I drifted off to sleep as soon as I closed my eyes are long gone. More usually, as my head hits the pillow, thoughts of papers or lectures I need to write, letters unwritten, patients to call, meetings to arrange, enter my mind. But occasionally I see my narcolepsy patients sitting in my clinic, one minute chatting about their condition, the next minute fast asleep. And when I put my younger daughter to bed, it is like a light switch suddenly turned off. Looking at her and my patients, it is obvious why we think of wake and sleep as entirely different states of existence, with clear borders: a reinforced concrete barrier, the Berlin Wall separating the East of sleep from the West of wake. But this seemingly stark divide belies the incredibly complex choreographed dance of brain nuclei, neurones and circuits, working synergistically and antagonistically to mediate our levels of engagement with our external and internal world that define our states of consciousness. Areas of the brain act in harmony to set our circadian rhythm, get us to sleep and guide us through the cycles of sleep. As we progress through the night, these circuits regulate the passage of sleep from light to deep non-REM sleep, then into REM sleep, four or five times.

As we all know, the more complex a system, the more likely it is to develop glitches. As I watch the computer engineers in our hospital wrestle with the nightmare of our IT systems whenever I have a technical issue, I realise that the days of switching the basic calculator of my school days off and on again when it stops working have truly passed. Our brains are infinitely more intricate than any man-made system, so it is perhaps surprising that such a complex system does not go awry more often than it does.

The range of sleep disorders is vast. From those who sleep too much, to those who sleep too little, and those who sleep at the wrong time or in the wrong way – these different sleep disorders allow us insights into the workings of the brain; how the brain influences sleep, but also how sleep influences the brain.

The functions and dysfunctions of the brain in wakefulness are myriad. In our daytime lives, we experience normal emotions, memories, cognition, everything that defines us as humans and individual conscious beings. And when these processes go wrong, we experience disorders, such as anxiety, depression, dementia, epilepsy, migraine, and so on. But while we think of our brains switching off at night, these patients illustrate that quite the opposite is true. The functions and dysfunctions of our nocturnal brains are as numerous and varied as those of the daytime, and influence every aspect of our waking lives too.

However, while the progress we have made in understanding the relationship between our nocturnal and daytime lives has been tremendous, I can't help but feel that we have only just scratched the surface. Many questions remain unanswered.

Some of these questions are almost inconceivably big, such as the true function (or functions) of dreaming, or whether we really prevent Alzheimer's disease by improving sleep. Others are less fundamental, but equally important to sufferers of a particular condition, like how precisely the immune system targets neurones in narcolepsy, what actually causes Kleine–Levin syndrome, or whether there is a cure for these conditions.

But we live in an era of great hope. As the disciplines of genetics, neuroscience and technology move forward, techniques to study sleep advance at an astounding pace. Our ability to track sleep, not only movement, at home for prolonged periods of time will help. So will methods to identify and analyse the genetics of individuals in huge numbers. And novel techniques of studying and influencing the brain, like using magnetic fields or electrical stimulation, will offer us new insights. Parallel to all this, the development of technological solutions to the collection and analysis of 'big data', for example the medical data and sleep parameters for huge populations, will be crucial.

And so I dream that many of these questions surrounding sleep will be answered in my lifetime. I look forward to the day when I sit in my clinic and, faced with a question from my patients, I no longer have to admit that I just don't know.

APPENDIX OF DIAGRAMS

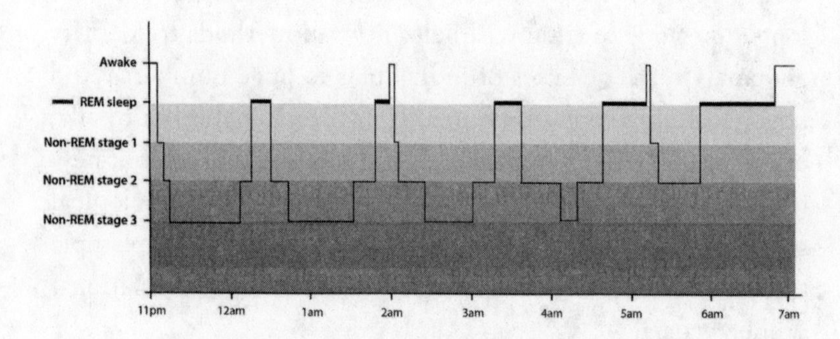

This hypnogram is a graphical representation of a typical night's sleep for a young adult. We cycle, on average, four to five times through the various stages of sleep, entering into REM sleep usually some 60–90 minutes after sleep onset. With each successive cycle, less non-REM sleep is seen, with increasing periods of REM sleep as we progress through the night. Brief awakenings are common and, when arising from REM sleep, as in this hypnogram, may result in the recollection of dreams.

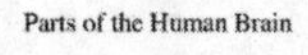
Parts of the Human Brain

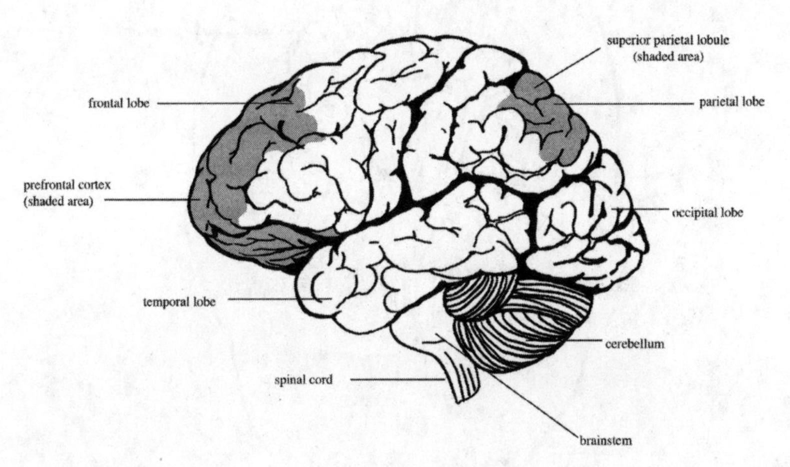

Major anatomical divisions of the brain. The frontal lobe has a wide array of functions, including control, initiation of movement, and, in the prefrontal cortex, planning, decision-making and regulation of behaviour. In sleepwalking (Chapters 2 and 10), the prefrontal cortex often exhibits reduced activity, explaining impaired reasoning and planning. The parietal lobe is the site of sensory processing, and, in particular, the superior parietal lobule is the site of representation of one's body in space (see Chapter 9).

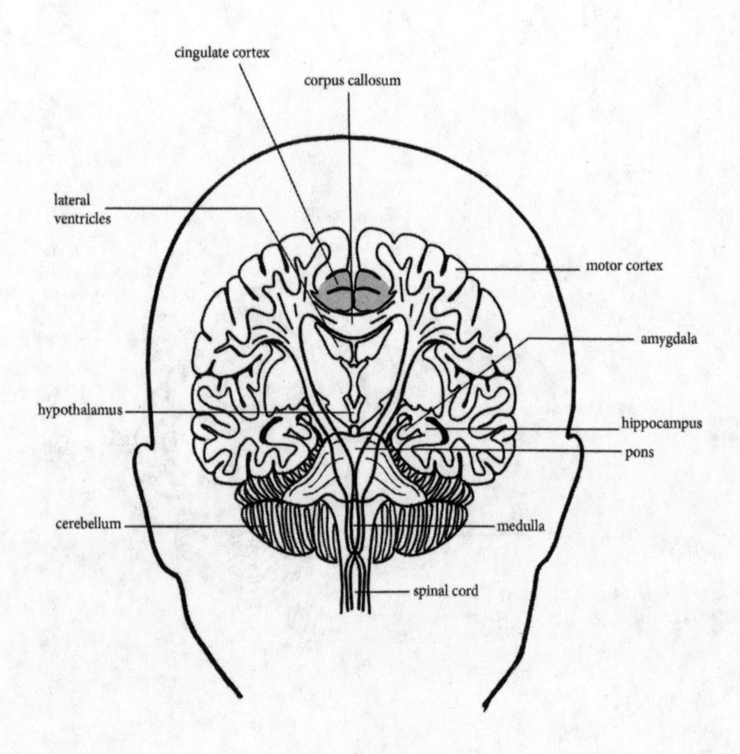

Cross-sectional cut of the brain. The cingulate cortex, hippocampus and amygdala all contribute to the limbic system, which mediates emotions, memories and arousal. Activation of this limbic system is fundamental to the 'flight-fight-fright' response, and plays an important role in sleepwalking (Chapters 2 and 10) and in some types of epilepsy (Chapter 8). Many of the important nuclei that regulate wake and sleep are located in the hypothalamus, pons and medulla.

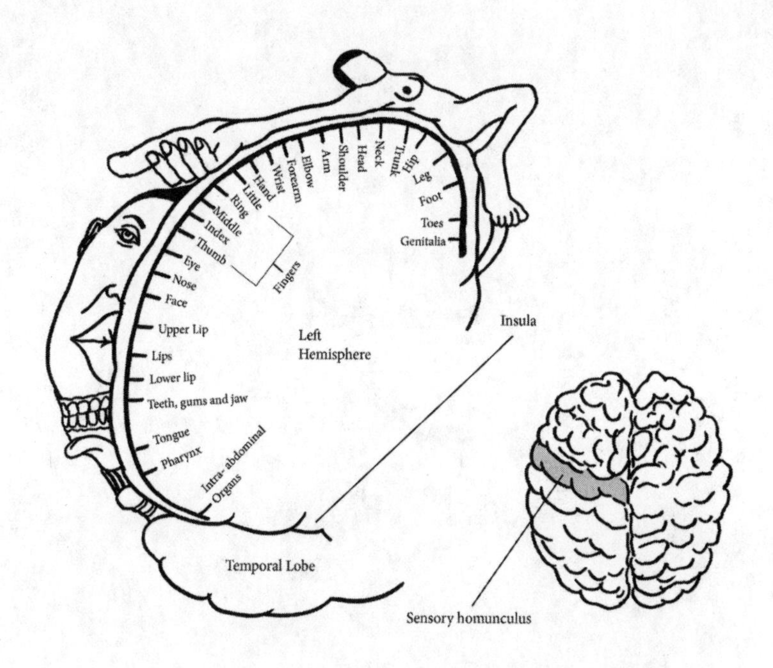

The sensory homunculus. Bodily sensations are processed by the primary sensory cortex, but the body is represented in a distorted way, with more sensitive body parts such as the face and hands represented by larger areas of cerebral cortex. The tongue, pharynx and intra-abdominal organs are represented in part of the sensory cortex overlying the insula, explaining some of the symptoms experienced in seizures arising from the insula (Chapter 8).

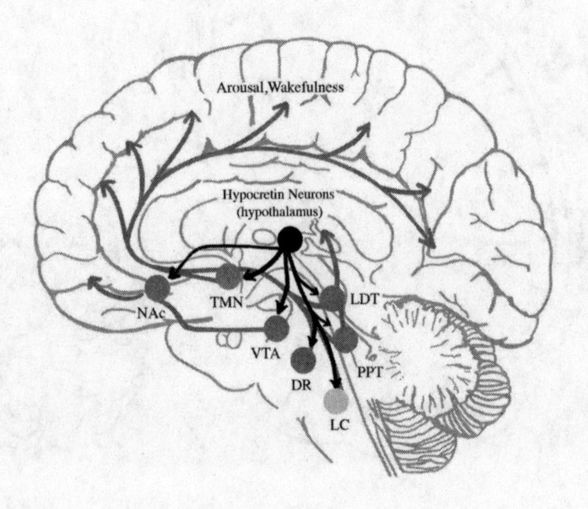

The hypocretin-producing neurones located in the hypothalamus project very widely, acting on several nuclei that promote wakefulness and non-REM sleep. These include the tuberomammillary nucleus (TMN), laterodorsal and pedunculopontine tegmental nuclei (LDT/PPT), dorsal raphe (DR), locus coeruleus (LC) and nucleus accumbens (NAc). Damage to the hypocretin neurones in narcolepsy (Chapters 6 and 13) results in destabilisation of these circuits, causing patients to flick in and out of sleep, especially REM sleep. This instability results in the hallmark features of this condition, such as sleep attacks, sleep paralysis, hallucinations and cataplexy.

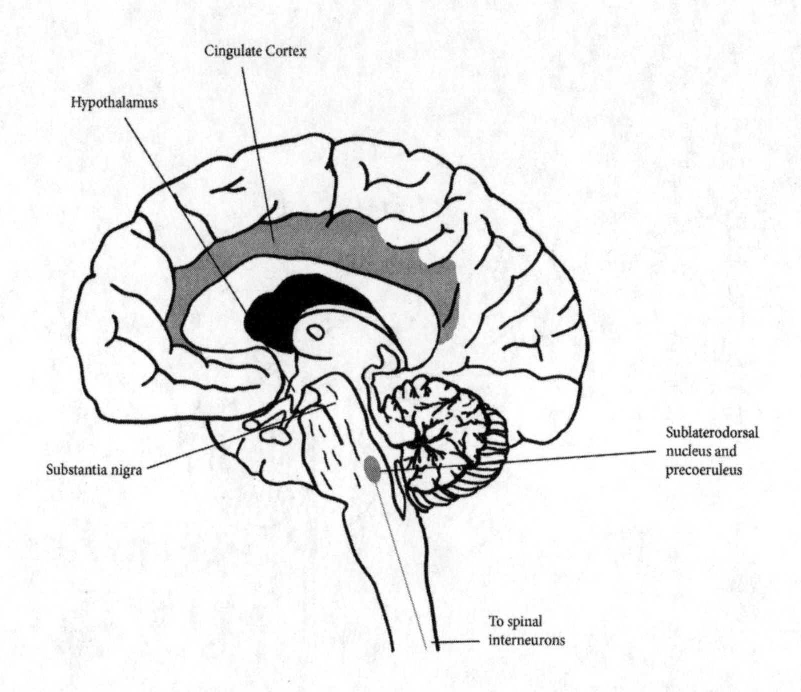

Cross-sectional view of the brain from the side, demonstrating the location of the cingulate cortex, part of the limbic system. The substantia nigra, the area of the brain that degenerates in Parkinson's disease, sits in close proximity to the hypothalamus. Nuclei in the pons called the sublaterodorsal nucleus and precoeruleus are active in REM sleep, and project to the spinal cord, normally inducing paralysis in REM sleep. Damage to this circuitry results in the loss of paralysis in REM sleep seen in REM sleep behaviour disorder (Chapter 3).

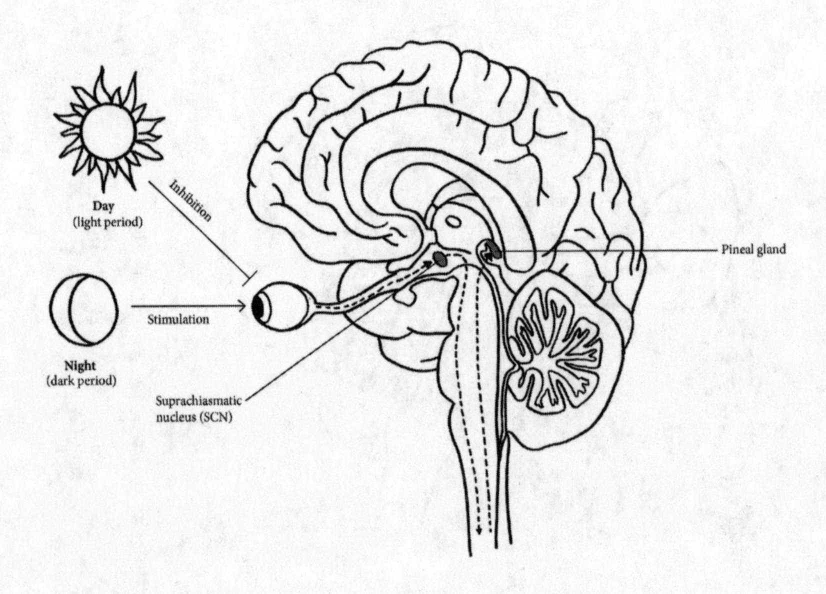

The suprachiasmatic nucleus (SCN) is the location of the 'master clock', which maintains the circadian rhythm (Chapter 1). Direct projections from the retina allow the SCN to be influenced by light, resulting in shifts in the circadian rhythm as a result of light exposure. The SCN mediates some of its effects by controlling the release of melatonin by the pineal gland.

ACKNOWLEDGEMENTS

First and foremost, this book would not exist without the help and kindness of the patients described, and their willingness to share their stories. They have been largely driven by their desire to highlight their experiences, and to help disseminate knowledge of their conditions, so that others experiencing similar problems might achieve a diagnosis and treatment more quickly. I am enormously grateful to them all. Equally, I have put together the amalgam of research in the field, presenting the hard and brilliant work of many researchers in the fields of sleep, neuroscience and clinical neurology. They are rapidly pushing at the boundaries of our understanding of sleep and the brain.

This book has come into fruition entirely accidentally. I had never anticipated writing anything other than academic texts. It is entirely due to my agent, Luigi Bonomi, who emailed me out of the blue, having heard the BBC Radio 4 series, *Mysteries of Sleep*. He convinced me to give writing a try. Luigi also succeeded in convincing Iain MacGregor, my superb editor at Simon & Schuster, to commission the book, who showed amazing enthusiasm for the project from the start. Thanks must also go to my agents in the US and Europe, George Lucas at InkWell and Nicki Kennedy at ILA, and Michael Flamini, my US editor at St Martin's Press.

Some of the work in this book originated from *Mysteries of Sleep*, and that is largely down to my fantastic producer, Sally Abrahams, who taught me how to tell a story, and the team at BBC Radio 4, who allowed me to make the series in the first place: Hugh Levinson, Mohit Bakaya and Richard Vadon. I am also grateful to my colleagues who contributed either to the radio series or through discussions regarding the contents of this book: Adrian Williams, Brian Kent, Ivana Rosenzweig, David O'Regan, Alex Nesbitt, Paul Gringras, Michael Farquhar, Sofia Eriksson, Sean Higgins, Mike Koutroumanidis, Al Santhouse, Russell Foster, Michael Kopelman, Annett Schrag and Meir Kryger. Meir also showed me that it is possible to bridge the gap between a busy clinical workload, research and writing. Thanks must also go to Allan Hobson, at Harvard, who patiently talked me through his theories about REM sleep.

My friends over the years have been bored rigid by my tales of neurology and sleep. Some have been further gluttons for punishment and have been volunteered to provide critical feedback – Jonathan Turner, Richard Ambrose and Rob Mills.

I am grateful to Guy's and St Thomas' NHS Trust, which has allowed us to develop a superb sleep centre (in my biased view unparalleled in the UK), staffed by a huge number of incredibly diligent, bright and capable individuals who work as a fantastic team. Also the excellent sleep team at London Bridge Hospital.

Finally, it is important to recognise my family. My parents, who triggered and encouraged my interest in science and enabled the pursuit of my medical career, and my children, Maya and Ava, who put up with me retreating into the study to write in spare moments. Finally, and most importantly, my wife Kavita, whose encouragement and critique has shaped much of this book. She always told me I had a book inside me, and I ignored her for years. As ever, I should have listened to her.

GLOSSARY

Amygdala – an almond-shaped structure deep in the temporal lobe, part of the network that constitutes the limbic system. It has a fundamental role in emotional responses such as fear, aggression and anxiety, but also contributes to memory and decision-making

Apnoea – a pause in breathing

Autonomic nervous system – a part of the nervous system that mediates largely unconscious control of the internal organs and skin. The sympathetic system drives the 'fright-fight-flight' response, causing an increase in heart rate, sweating, pupil dilatation and diversion of blood away from the gut and skin to the muscles, heart and lungs. The parasympathetic system mediates the opposite response: that associated with being relaxed

Cataplexy – a phenomenon found almost exclusively in narcolepsy. Strong emotions, particularly laughter, result in sudden loss of muscle strength, causing transient weakness in various body parts or even the whole body

Cerebral cortex – the outer layer of the brain, also known as grey matter

Chronotype – the tendency for an individual to fall asleep and wake up at a particular time, e.g. a 'morning person' or an 'evening person'

Cingulate – an area of the brain, the cortex of which plays an integral role in the limbic system, the network of the brain responsible for emotion, behaviour and motivation

Circadian – a biological process recurring on a 24-hour cycle

Delayed sleep phase syndrome (or disorder) – a shift of the internal body clock forward, resulting in sufferers only being able to go to sleep much later, and wanting to wake up much later, to such an extent that it has a negative impact on their waking lives

Dementia with Lewy bodies – a degenerative condition of the brain, resulting in cognitive dysfunction and hallucinations. It has significant overlap with Parkinson's disease, both in terms of symptoms and signs but also on a microscopic level

EEG (electroencephalogram) – a technique utilised to study electrical activity of the brain. Different appearances on the electrical traces permit the differentiation of different stages of sleep, and abnormal brain activity caused by epilepsy or other neurological disorders

Frontal lobe – the region of the brain closest to the forehead, above the eye sockets. Its functions include control of voluntary movement, planning, judgement, decision-making and emotional expression

Homeostatic mechanism – one of the processes that regulates sleep. The longer you are awake, the stronger the drive to enter sleep

Hypnagogic hallucinations – hallucinations, often in the form of an intruder in the room or an out-of-body experience, at the point of drifting off to sleep or waking (termed hypnopompic). They are thought to represent intrusion of dreamlike processes into wakefulness

Hypocretin – the neurotransmitter that is lost in narcolepsy as a result of the death of or damage to the neurones producing this chemical

Hypothalamus – a small area of the brain behind and between the eyes containing multiple nuclei important in the regulation of metabolic processes, hunger, sleep, circadian rhythms, thirst and body temperature

Insula – an area of cerebral cortex that is covered by the parietal, frontal and temporal lobes. It acts as a junction between these parts of the brain and the limbic system

Lesion – the term for an area of damage to tissue. In neurological terms, it defines a location in the nervous system that is diseased, damaged or in some way not functioning correctly

Limbic system – a brain network comprising a set of structures including the hippocampus, amygdala, cingulate gyrus, thalamus, fundamental to the integration of memories, emotions and smells. It influences motivation, emotional experiences and behaviour

Localisation – part of the diagnostic process focused on identifying the location of a lesion within the nervous system

Lucid dreaming – the maintenance of a degree of consciousness or conscious control during dreaming

Narcolepsy – a neurological disorder that results in the inability to regulate sleep and dreaming, thought to be caused by the destruction of brain cells producing hypocretin, a neurotransmitter fundamental to the control of sleep, in the hypothalamus. It results in excessive sleepiness, hypnagogic hallucinations, sleep paralysis and cataplexy (see Chapters 6 and 13)

Non-REM parasomnias – conditions causing abnormal behaviours such as sleepwalking, night terrors, sleep-talking, or sleep-eating, all arising from non-REM sleep, particularly Stage 3 sleep

Non-REM sleep – comprises Stage 1, Stage 2 and Stage 3 sleep

Non-24-hour rhythm disorder – a syndrome defined by the circadian clock running on a rhythm that is outside the normal 24-hour interval, usually longer. Also known as free-running disorder

Parasomnia – any abnormal behaviour arising from sleep

Parietal lobe – the area of the brain principally involved in sensation, representation of the physical world around us, and our own bodies within the world

Parkinson's disease – a common disorder of brain degeneration, principally affecting movement, resulting in tremor, difficulty walking, stiffness and slowing

Pineal gland – a small, cone-shaped structure a few millimetres in diameter that secretes melatonin. It is located deep within the brain, behind one of the fluid cavities known as the third ventricle, as it is itself bathed in cerebrospinal fluid

Prefrontal cortex – an area of cerebral cortex in the frontal lobe, implicated in the making of decisions, planning of actions, social behaviour and expression of personality

REM (rapid eye movement) sleep – the stage of sleep most associated with dreaming, characterised by rapid movements of the eyes from side to side, paralysis of almost all muscles, but an active brain state

REM sleep behaviour disorder – loss of paralysis of muscles in REM sleep, resulting in the acting out of dreams

Retinal ganglion cells – light-detecting cells in the retina that have no function in vision. They detect blue light in particular and project directly to the suprachiasmatic nucleus via the retinohypothalamic tract, feeding information about environmental light to the master clock in the suprachiasmatic nucleus

Sexsomnia – a form of non-REM parasomnia resulting in behaviour of a sexual nature

Sleep apnoea – the recurrent pausing of breathing in sleep, usually related to partial or complete obstruction of the airway as it becomes more floppy in sleep

Sleepwalking – a form of non-REM parasomnia. Behaviours arising from Stage 3 non-REM sleep, consisting of getting out of bed, performing complex tasks and interacting with the environment, usually with no or very limited recall

Slow wave sleep – *see* Stage 3 sleep

Stage 1 sleep – the lightest stage of sleep, also known as drowsiness. It is defined by slow, rolling eye movements and a quietening of electrical activity of the brain when measured on the EEG

Stage 2 sleep – intermediate sleep, defined by signature patterns of the EEG, known as sleep spindles and K-complexes

Stage 3 sleep – the deepest stage of sleep, also known as slow wave sleep, thought to be most important for restoration of function and recovery. The brainwaves become much slower and larger and in this stage of sleep it is the most difficult to rouse someone

Suprachiasmatic nucleus – the tiny area of the hypothalamus that represents the master clock of the body, maintaining the circadian rhythm

Temporal lobe – a region of the brain responsible for language function, hearing and auditory processing. The inner aspect of the temporal lobe contains the hippocampus and amygdala, and is therefore also involved in emotional processing and memory

Zeitgeber – a factor external to the master clock in the suprachiasmatic nucleus that influences the circadian rhythm, such as light and melatonin

FURTHER READING

It is impossible to include all references without adding a further 100 pages to this book. Below is a list of key publications related to the chapters, largely review articles. Where the cases have been published in the medical literature, these references have also been included.

Chapter 1: Greenwich Mean Time

Malkani, R. G., Abbott, S. M., Reid, K. J., Zee, P. C., 'Diagnostic and Treatment Challenges of Sighted Non-24-Hour Sleep–Wake Disorder', *J Clin Sleep Med*, 15 April 2008, 14(4): 603–613.

Quera Salva, M. A., Hartley, S., Léger, D., Dauvilliers, Y. A., 'Non-24-Hour Sleep–Wake Rhythm Disorder in the Totally Blind: Diagnosis and Management', *Front Neurol*, 18 December 2017, 18(8): 686.

Uchiyama, M., Lockley, S. W., 'Non-24-Hour Sleep–Wake Rhythm Disorder in Sighted and Blind Patients', *Sleep Med Clin*, December 2015, 10(4): 495–516.

Hayakawa, T., Uchiyama, M., Kamei, Y., Shibui, K., Tagaya, H., Asada, T., Okawa, M., Urata, J., Takahashi, K., 'Clinical analyses of sighted patients with non-24-hour sleep–wake syndrome: a study of 57 consecutively diagnosed cases', *Sleep*, 1 August 2005, 28(8): 945–52.

Edgar, R. S., Green, E. W., Zhao, Y., van Ooijen, G., Olmedo, M., Qin, X., Xu, Y., Pan, M., Valekunja, U. K., Feeney, K. A., Maywood, E. S., Hastings, M. H., Baliga, N. S., Merrow, M., Millar, A. J., Johnson, C. H., Kyriacou, C. P., O'Neill, J. S., Reddy, A. B., 'Peroxiredoxins are conserved markers of circadian rhythms', *Nature*, May 2012, 485(7399): 459–64.

Jagannath, A., Taylor, L., Wakaf, Z., Vasudevan, S. R., Foster, R. G., 'The genetics of circadian rhythms, sleep and health', *Hum Mol Genet*, 1 October 2017, 26(R2): R128–R138.

Touitou, Y., Reinberg, A., Touitou, D., 'Association between light at night, melatonin secretion, sleep deprivation, and the internal clock: Health impacts and mechanisms of circadian disruption', *Life Sci*, 15 March 2017, 173: 94–106. doi: 10.1016/j.lfs.2017.02.008.

Travis, R. C., Balkwill, A., Fensom, G. K., Appleby, P. N., Reeves, G. K., Wang, X. S., Roddam, A. W., Gathani, T., Peto, R., Green, J., Key, T. J., Beral, V., 'Night Shift Work and Breast Cancer Incidence: Three Prospective Studies and Meta-analysis of Published Studies', *J Natl Cancer Inst*, 6 October 2016, 108(12).

Chapter 2: In the Still of the Night

Bargiotas, P., Arnet, I., Frei, M., Baumann, C. R., Schindler, K., Bassetti, C. L., 'Demographic, Clinical and Polysomnographic Characteristics of Childhood- and Adult-Onset Sleepwalking in Adults', *Eur Neurol*, 2017, 78(5–6): 307–11.

Bassetti, C., Vella, S., Donati, F., Wielepp, P., Weder, B., 'SPECT during sleepwalking', *Lancet*, 5 August 2000, 356(9228): 484–5.

Drakatos, P., Marples, L., Muza, R., Higgins, S., Gildeh, N., Macavei, R., Dongol, E. M., Nesbitt, A., Rosenzweig, I., Lyons, E., d'Ancona, G., Steier, J., Williams, A. J., Kent, B. D., Leschziner, G., 'NREM parasomnias: a treatment approach based upon a retrospective case series of 512 patients', *Sleep Med*, 10 April 2018. pii: S1389-9457 (18)30099-6.

Iranzo, A., 'Parasomnias and Sleep-Related Movement Disorders in Older Adults', *Sleep Med Clin*, March 2018, 13(1): 51–61.

Pressman, M. R., 'Factors that predispose, prime and precipitate NREM parasomnias in adults: clinical and forensic implications', *Sleep Med Rev*, February 2007, 11(1): 5–3.

Moreno, M. A., 'Sleep Terrors and Sleepwalking: Common Parasomnias of Childhood', *JAMA Pediatr*, July 2015, 169(7): 704.

Chapter 3: Disney Was Right

Oudiette, D., De Cock, V. C., Lavault, S., Leu, S., Vidailhet, M., Arnulf, I., 'Nonviolent elaborate behaviors may also occur in REM sleep behavior disorder', *Neurology*, 10 February 2009, 72(6): 551–7.

Aserinsky, E., 'The discovery of REM sleep', *J Hist Neurosci*, December 1996, 5(3): 213–27.

Iranzo, A., Stefani, A., Serradell, M., Martí, M. J., Lomeña, F., Mahlknecht, P., Stockner, H., Gaig, C., Fernández-Arcos, A., Poewe, W., Tolosa, E., Högl, B., Santamaria, J., 'Characterization of patients with longstanding idiopathic REM sleep behavior disorder', SINBAR (Sleep Innsbruck Barcelona) group, *Neurology*, 18 July 2017, 89(3): 242–8.

Postuma, R. B., Iranzo, A., Hogl, B., Arnulf, I., Ferini-Strambi, L., Manni, R., Miyamoto, T., Oertel, W., Dauvilliers, Y., Ju,

Y. E., Puligheddu, M., Sonka, K., Pelletier, A., Santamaria, J., Frauscher, B., Leu-Semenescu, S., Zucconi, M., Terzaghi, M., Miyamoto, M., Unger, M. M., Carlander, B., Fantini, M. L., Montplaisir, J. Y., 'Risk factors for neurodegeneration in idiopathic rapid eye movement sleep behavior disorder: a multicenter study', *Ann Neurol*, May 2015, 77(5): 830–9.

Boeve, B. F., Silber, M. H., Ferman, T. J., Lin, S. C., Benarroch, E. E., Schmeichel, A. M., Ahlskog, J. E., Caselli, R. J., Jacobson, S., Sabbagh, M., Adler, C., Woodruff, B., Beach, T. G., Iranzo, A., Gelpi, E., Santamaria, J., Tolosa, E., Singer, C., Mash, D. C., Luca, C., Arnulf, I., Duyckaerts, C., Schenck, C. H., Mahowald, M. W., Dauvilliers, Y., Graff-Radford, N. R., Wszolek, Z. K., Parisi, J. E., Dugger, B., Murray, M. E., Dickson, D. W., 'Clinicopathologic correlations in 172 cases of rapid eye movement sleep behavior disorder with or without a coexisting neurologic disorder', *Sleep Med*, August 2013, 14(8): 754–62.

Chapter 4: Rumblings

Polsek, D., Gildeh, N., Cash, D., Winsky-Sommerer, R., Williams, S. C. R., Turkheimer, F., Leschziner, G. D., Morrell, M. J., Rosenzweig, I., 'Obstructive sleep apnoea and Alzheimer's disease: In search of shared pathomechanisms', *Neurosci Biobehav Rev*, 7 December 2017. pii: S0149-7634(17)30435-9.

Hopps, E., Caimi, G., 'Obstructive Sleep Apnea Syndrome: Links Between Pathophysiology and Cardiovascular Complications', *Clin Invest Med*, 4 December 2015, 38(6): E362–70.

Emamian, F., Khazaie, H., Tahmasian, M., Leschziner, G. D., Morrell, M. J., Hsiung, G. Y., Rosenzweig, I., Sepehry,

A. A., 'The Association Between Obstructive Sleep Apnea and Alzheimer's Disease: A Meta-Analysis Perspective', *Front Aging Neurosci*, 12 April 2016, 8(78).

Yu, J., Zhou, Z., McEvoy, R. D., Anderson, C. S., Rodgers, A., Perkovic, V., Neal, B., 'Association of Positive Airway Pressure With Cardiovascular Events and Death in Adults With Sleep Apnea: A Systematic Review and Meta-analysis', *JAMA*, 11 July 2017, 318(2): 156–66.

Abuzaid, A. S., Al Ashry, H. S., Elbadawi, A., Ld, H., Saad, M., Elgendy, I. Y., Elgendy, A., Mahmoud, A. N., Mentias, A., Barakat, A., Lal, C., 'Meta-Analysis of Cardiovascular Outcomes With Continuous Positive Airway Pressure Therapy in Patients With Obstructive Sleep Apnea', *Am J Cardiol*, 15 August 2017, 120(4): 693–9.

Javaheri, S., Barbe, F., Campos-Rodriguez, F., Dempsey, J. A., Khayat, R., Javaheri, S., Malhotra, A., Martinez-Garcia, M. A., Mehra, R., Pack, A. I., Polotsky, V. Y., Redline, S., Somers, V. K., 'Sleep Apnea: Types, Mechanisms, and Clinical Cardiovascular Consequences', *J Am Coll Cardiol*, 21 February 2017, 69(7): 841–58.

Chapter 5: The Sleep-Talking Bus Driver

Bashford, J., Leschziner, G., 'Bed Partner "Gas-Lighting" as a cause of fictitious sleep-talking', *J Clin Sleep Med*, 15 October 2015, 11(10): 1237–8.

Chapter 6: Weak With Laughter

Leschziner, G., 'Narcolepsy: a clinical review', *Practical Neurology*, October 2014, 14(5): 323–31.

Overeem, S., Lammers, G. J., van Dijk, J. G., 'Cataplexy: "tonic immobility" rather than "REM-sleep atonia"?', *Sleep Med*, November 2002, 3(6): 471–7.

Sarkanen, T., Alakuijala, A., Julkunen, I., Partinen, M., 'Narcolepsy Associated with Pandemrix Vaccine', *Curr Neurol Neurosci Rep*, 1 June 2018, 18(7): 43.

Sturzenegger, C., Bassetti, C. L., 'The clinical spectrum of narcolepsy with cataplexy: a reappraisal', *J Sleep Res*, December 2004, 13(4): 395–406.

Stowe, J., Miller, E., Andrews, N., Kosky, C., Leschziner, G., Shneerson, J. M., Hall, A., Eriksson, S., Reading, P., Dennis, G., Donegan, K., 'Risk of Narcolepsy after AS03 Adjuvanted Pandemic A/H1N1 2009 Influenza Vaccine in Adults: A Case-Coverage Study in England', *Sleep*, 1 May 2016, 39(5): 1051–7.

Drakatos, P., Leschziner, G., 'Cataplexy with Normal Sleep Studies and Normal CSF Hypocretin: an Explanation?', *J Clin Sleep Med*, 15 March 2016, 12(3): 449–50.

Chapter 7: Buzzing Bees

Leschziner, G., Gringras, P., 'Restless Legs Syndrome', *British Medical Journal*, 23 May 2012, 344: e3056.

Athauda, D., Leschziner, G., 'A restless night's sleep', *British Medical Journal*, 2012, 344: d8347.

Schormair, B., Zhao, C., Bell, S., Tilch, E., Salminen, A. V., Pütz, B., Dauvilliers, Y., Stefani, A., Högl, B., Poewe, W., Kemlink, D., Sonka, K., Bachmann, C. G., Paulus, W., Trenkwalder, C., Oertel, W. H., Hornyak, M., Teder-Laving, M., Metspalu, A., Hadjigeorgiou, G. M., Polo, O., Fietze, I., Ross, O. A., Wszolek, Z., Butterworth, A. S., Soranzo, N., Ouwehand, W. H., Roberts, D. J., Danesh, J., Allen, R. P., Earley, C. J., Ondo, W. G., Xiong,

L., Montplaisir, J., Gan-Or, Z., Perola, M., Vodicka, P., Dina, C., Franke, A., Tittmann, L., Stewart, A. F. R., Shah, S. H., Gieger, C., Peters, A., Rouleau, G. A., Berger, K., Oexle, K., Di Angelantonio, E., Hinds, D. A., Müller-Myhsok, B., Winkelmann, J., 'Identification of novel risk loci for restless legs syndrome in genome-wide association studies in individuals of European ancestry: a meta-analysis', 23andMe Research Team, DESIR study group, *Lancet Neurol*, November 2017, 16(11): 898–907. doi: 10.1016/S1474-4422(17)30327-7. Review.

Winkelmann, J., Allen, R. P., Högl, B., Inoue, Y., Oertel, W., Salminen, A. V., Winkelman, J. W., Trenkwalder, C., Sampaio, C., 'Treatment of restless legs syndrome: Evidence-based review and implications for clinical practice (Revised 2017)', *Mov Disord*, 14 May 2018. doi: 10.1002/mds.27260.

Chapter 8: Seized by the Throat

Schindler, K., Gast, H., Bassetti, C., Wiest, R., Fritschi, J., Meyer, K., Kollar, M., Wissmeyer, M., Lövblad, K., Weder, B., Donati, F., 'Hyperperfusion of anterior cingulate gyrus in a case of paroxysmal nocturnal dystonia', *Neurology*, 11 September 2001, 57(5): 917–20.

Nesbitt, A., Kosky, C. A., Leschziner, G. D., 'Insular seizures causing sleep-related breathlessness', *The Lancet*, 2013, 382: 1756.

Tinuper, P., Bisulli, F., 'From nocturnal frontal lobe epilepsy to Sleep-Related Hypermotor Epilepsy: A 35-year diagnostic challenge', January 2017, 44: 87–92. doi: 10.1016/j.seizure.2016.11.023.

Derry, C. P., 'Sleeping in fits and starts: a practical guide to distinguishing nocturnal epilepsy from sleep disorders', *Pract Neurol*, December 2014, 14(6): 391–8.

Nobili, L., Proserpio, P., Combi, R., Provini, F., Plazzi, G., Bisulli, F., Tassi, L., Tinuper, P., 'Nocturnal frontal lobe epilepsy', *Curr Neurol Neurosci Rep*, February 2014, 14(2): 424.

Chapter 9: Floating Eyeballs

Jalal, B., Ramachandran, V. S., 'Sleep Paralysis, "The Ghostly Bedroom Intruder" and Out-of-Body Experiences: The Role of Mirror Neurons', *Front Hum Neurosci*, 28 February 2017, 11: 92.

Jalal, B., Ramachandran, V. S., 'Sleep paralysis and "the bedroom intruder": the role of the right superior parietal, phantom pain and body image projection', *Med Hypotheses*, December 2014, 83(6): 755–7.

Denis, D., French, C. C., Gregory, A. M., 'A systematic review of variables associated with sleep paralysis', *Sleep Med Rev*, April 2018, 38: 141–57.

Molendijk, M. L., Montagne, H., Bouachmir, O., Alper, Z., Bervoets, J. P., Blom, J. D., 'Prevalence Rates of the Incubus Phenomenon: A Systematic Review and Meta-Analysis', *Front Psychiatry*, 24 November 2017, 8: 253.

Sharpless, B. A., 'A clinician's guide to recurrent isolated sleep paralysis', *Neuropsychiatr Dis Treat*, 19 July 2016, 12: 1761–7.

Chapter 10: Jekyll and Hyde

Siclari, F., Khatami, R., Urbaniok, F., Nobili, L., Mahowald, M. W., Schenck, C. H., Cramer Bornemann, M. A., Bassetti, C. L., 'Violence in sleep', *Brain*, December 2010, 133(Pt 12): 3494–509.

Dubessy, A. L., Leu-Semenescu, S., Attali, V., Maranci, J. B., Arnulf, I., 'Sexsomnia: A Specialized Non-REM Parasomnia?', *Sleep*, 1 February 2017, 40(2).

Pressman, M. R., Mahowald, M. W., Schenck, C. H., Cramer Bornemann, M. A., Banerjee, D., Buchanan, P., Zadra, A., 'Alcohol, sleepwalking and violence: lack of reliable scientific evidence', *Brain*, February 2013, 136(Pt 2): e229.

Morrison, I., Rumbold, J. M., Riha, R. L., 'Medicolegal aspects of complex behaviours arising from the sleep period: a review and guide for the practising sleep physician', *Sleep Med Rev*, June 2014, 18(3): 249–60. doi: 10.1016/j.smrv.2013.07.004.

Chapter 11: The Waking Effects of Coffee

Inoue, Y., 'Sleep-related eating disorder and its associated conditions', *Psychiatry Clin Neurosci*, June 2015, 69(6): 309–20.

Vander Wal, J. S., 'Night eating syndrome: a critical review of the literature', *Clin Psychol Rev*, February 2012, 32(1): 49–59.

Howell, M. J., 'Restless Eating, Restless Legs, and Sleep Related Eating Disorder', *Curr Obes Rep*, March 2014, 3(1): 108–13.

Howell, M. J., Schenck, C. H., 'Restless nocturnal eating: a common feature of Willis–Ekbom Syndrome (RLS)', *J Clin Sleep Med*, 15 August 2012, 8(4): 413–9.

Chapter 12: A Peculiar Fairy Tale

Nesbitt, A., Leschziner, G., 'Migraine with brainstem aura presenting as recurrent hypersomnia (Kleine–Levin Syndrome)', *Practical Neurology*, October 2016, 16(5): 402–5.

Gadoth, N., Oksenberg, A., 'Kleine–Levin syndrome; An update and mini-review', *Brain Dev*, September 2017, 39(8): 665–71.

Miglis, M. G., Guilleminault, C., 'Kleine–Levin Syndrome', *Curr Neurol Neurosci Rep*, June 2016, 16(6): 60.

Lavault, S., Golmard, J. L., Groos, E., Brion, A., Dauvilliers, Y., Lecendreux, M., Franco, P., Arnulf, I., 'Kleine–Levin syndrome in 120 patients: differential diagnosis and long episodes', *Ann Neurol*, March 2015, 77(3): 529–40.

Chapter 13: Inception

Hobson, J. A., Hong, C. C., Friston, K. J., 'Virtual reality and consciousness inference in dreaming', *Front Psychol*, 9 October 2014, 5: 1133.

Hobson, J. A., 'REM sleep and dreaming: towards a theory of protoconsciousness', *Nat Rev Neurosci*, November 2009, 10(11): 803–13.

Voss, U., Holzmann, R., Tuin, I., Hobson, J. A., 'Lucid dreaming: a state of consciousness with features of both waking and non-lucid dreaming', *Sleep*, September 2009, 32(9): 1191–200.

Crick, F., Mitchison, G., 'The function of dream sleep', *Nature*, 14–20 July 1983, 304(5922): 111–4.

Cipolli, C., Ferrara, M., De Gennaro, L., Plazzi, G., 'Beyond the neuropsychology of dreaming: Insights into the neural basis of dreaming with new techniques of sleep recording and analysis', *Sleep Med Rev*, October 2017, 35: 8–20.

Dodet, P., Chavez, M., Leu-Semenescu, S., Golmard, J. L., Arnulf, I., 'Lucid dreaming in narcolepsy', *Sleep*, 1 March 2015, 38(3): 487–97.

Dresler, M., Koch, S. P., Wehrle, R., Spoormaker, V. I., Holsboer, F., Steiger, A., Sämann, P. G., Obrig, H., Czisch, M.,

'Dreamed movement elicits activation in the sensorimotor cortex', *Curr Biol*, 8 November 2011, 21(21): 1833–7.

van der Helm, E., Yao, J., Dutt, S., Rao, V., Saletin, J. M., Walker, M. P., 'REM sleep depotentiates amygdala activity to previous emotional experiences', *Curr Biol*, 6 December 2011, 21(23): 2029–32.

Chapter 14: Losing Sleep

Xie, L., Kang, H., Xu, Q., Chen, M. J., Liao, Y., Thiyagarajan, M., O'Donnell, J., Christensen, D. J., Nicholson, C., Iliff, J. J., Takano, T., Deane, R., Nedergaard, M., 'Sleep drives metabolite clearance from the adult brain', *Science*, 19 October 2013, 342(6156): 373–7.

Fernandez-Mendoza, J., Shea, S., Vgontzas, A. N., Calhoun, S. L., Liao, D., Bixler, E. O., 'Insomnia and incident depression: role of objective sleep duration and natural history', *J Sleep Res*, August 2015, 24(4): 390–98.

Li, Y., Vgontzas, A. N., Fernandez-Mendoza, J., Bixler, E. O., Sun, Y., Zhou, J., Ren, R., Li, T., Tang, X., 'Insomnia with physiological hyperarousal is associated with hypertension', *Hypertension*, March 2015, 65(3): 644–50.

Vgontzas, A. N., Fernandez-Mendoza, J., Liao, D., Bixler, E. O., 'Insomnia with objective short sleep duration: the most biologically severe phenotype of the disorder', *Sleep Med Rev*, August 2013, 17(4): 241–54.

Mitchell, M. D., Gehrman, P., Perlis, M., Umscheid, C. A., 'Comparative effectiveness of cognitive behavioral therapy for insomnia: a systematic review', *BMC Fam Pract*, 25 May 2012, 13: 40.

Jarrin, D. C., Alvaro, P. K., Bouchard, M. A., Jarrin, S. D., Drake, C. L., Morin, C. M., 'Insomnia and hypertension:

A systematic review', *Sleep Med Rev.*, 16 February 2018. pii: S1087-0792(17)30051-5.

Penninkilampi, R., Eslick, G. D., 'A Systematic Review and Meta-Analysis of the Risk of Dementia Associated with Benzodiazepine Use, After Controlling for Protopathic Bias', *CNS Drugs*, 20 June 2018. doi: 10.1007/s40263-018-0535-3.

INDEX

actigraphy, 26–7
Adrian (cataplexy), 111–15, 118–19, 121, 122, 125–6, 127, 130, 133, 261, 262
Alex (sleepwalking), 42–7, 56–7, 102, 200, 217
  anecdotes concerning, 43–4
  relaxed nature of, 43, 57
  treatments for, 56
alphasynuclein, 68
Alzheimer's disease, 86, 87–8, 92, 302
amnesia, 202, 242 (*see also* sleep)
amygdala, 123–4
ANP, 83
antioxidant mechanisms, 84
Aserinsky, Eugene, 62–4, 119
Asperger's, 198, 213
Athenaeus, 82
augmentation, 152–3
automatism, 211–12
Awakenings (Sacks), 116

Babinski, Joseph, 96
basal ganglia, 68, 73, 235
benzodiazepines, 56, 74, 301
Bergman, Ingrid, 107
Bethlem Hospital, 167
blood pressure, 64, 69, 83–4, 85, 91, 177, 292, 293
body clock, *see* circadian clock
brainstem, 11, 65, 67–8
  changes in, 71
  and H1N1, 128
  and protein deposits, 68
British Medical Journal, 144, 309
Brouillet, Pierre, 95–6
Brownies, 37

cancers, 30–1
Cartesian dualism, 273, 274, 280
cataplexy, 111–38 (*see also* Adrian; Phil), 114–15, 116, 118, 121–3, 124, 126, 130–1, 132–7
  laughter's association with, 123, 124, 132, 139
  laughter triggers, 122
  REM linked with, 122
Catholic Inquisition, 287

cerebral cortex, 52, 165, 170, 171–2
  superior parietal lobule in, 191
channelopathy, 252, 253
Charcot, Jean-Martin, 95–6, 110, 208
Charles Bonnet syndrome, 188–9
Christian (narcolepsy), 259–64, 271–3, 280–1, 283–5
Cinderella, 58, 65, 75
cingulate cortex, 40–1, 47
circadian clock, 18–20 (*see also* sleep)
  and all forms of life, 18–19
  and blindness, 27
  body's 'control room' for, 21
  and cancers, 30–1
  and DNA, 31
  genes' influence on, 11, 23
  and 'givers of time' (*Zeitgebers*), 19, 22, 27, 29
  and Great Oxygenation Event, 19–20
  hardwired into life, 19
  and knitted lives, 35
  and life's phases, 23–4
  and melatonin, 27–8, 33
  and school times, 30
  and shift work, 30, 31, 34–5
  and sleep phases, 17, 29, 32
  and suprachiasmatic nucleus, 21–2, 27
  thwarted attempt to change, 18
Claire (insomnia), 288–90, 296, 297–9, 303, 305–7
A Clinical Lesson at the Salpêtrière (Brouillet), 96
cognitive behavioural therapy (CBTi), 302–3, 304–6
continuous positive airway pressure (CPAP), 55, 89–93, 104–5, 308–9
cortisol, 291, 294
Creutzfeldt–Jakob (mad-cow) disease, 296
Crick, Francis, 278
  DNA co-discovered by, 2

Dahlia (Vincent's mother), 15–17, 25–6, 32
David (RLS), 142–4, 149–50, 157, 158–60
Debra (wife of David), 142, 143, 144, 150, 158–9
Dement, William, 63–4
Descartes, René, 27
diabetes, 30, 31, 84, 85
DiCaprio, Leonardo, 284
Dickens, Charles, 81
Dionysius, 82
Don (sleep-eating), 217–22, 223–30
  and psychological factors, 228

types of food eaten by, 221–2, 223
dopamine, 136, 152, 153, 155–6, 158, 163, 227, 229
and iron, 156
dreams (*see also* sleep):
of animal attacks, 60–1, 67, 71
and brain lesions, 11
and dreamer's model of world, 279
in *Egyptian Dream Book*, 265–6
emotional content of, 73
Freud's view of, *see* Freud, Sigmund
function of, 2, 275–6, 315
and jerky eye movements, 63
and lesions, 11
lucid, 193, 280–2
many forms of, 191
as narrative structure, 61–2
and night terrors, 59–61, 63, 65, 66 (*see also* John)
and paralysis, 187
and perceived mind–body rift, 273
in religious texts, 266
REM, *see* sleep: non-REM; sleep: REM
and sleep behaviour disorder, *see* REM sleep behaviour disorder
and sleep deprivation, 286
and sleep paralysis, 193
stories developing over time, 66
subjective, 268
and suppressed emotions, desires, 273
on waking from REM, 120
while awake, 119, 188

Ed (Jackie's partner), 50–1
EEG (electroencephalogram), 164–5, 173–4, 178, 179–81, 203, 208, 243, 267, 281, 309
Egyptian Dream Book, 265–6, 273, 308
Ekbom, Karl-Axel, 145, 155
electroencephalogram (EEG), 164–5, 173–4, 178, 179–81, 203, 208, 243, 267, 281, 309
encephalitis lethargica, 115–16
epilepsy, 105, 111, 171–82 *passim*, 202, 211
and amygdala, 123
brain onset of, 172
and brain surgery, 172
and electrical impulses, 169–70
familial forms of, 253
frequent description of, 169
frontal-lobe, 175
and insular cortex, 177–8
and music, 7
and religious experiences, 171
and sex hormones, 173
and sleep apnoea, 86
sleep's links with, 173–4

epilepsy – *continued*
  and sleepwalking, 41–2
  and stress, 97
  and sudden emotion, 123
  unusual forms of, 242
  various manifestations of, 171
Erin (Adrian's daughter), 113–14
Evelyn (hallucinations), 184–6, 187, 189, 190, 194–7, 217, 261, 280

Frankenstein (Shelley), 270
Freud, Sigmund, 2, 61, 96, 110, 217, 273–4, 279, 308
fright-fight-flight, 45, 47, 83, 276, 292, 299
Friston, Karl, 277–9
frontal lobes, 174–6
  and social behaviour, 10
Fuseli, Henry, 186–7

GABA, 136
Gage, Phineas, 8–10, 209, 210
Gareth (Alex's housemate), 44, 46
Gaslight, 107
gaslighting, 107
GBH, 136
genes, and chronotype, 23
glymphatic system, 87
Gowers Round, 4–5, 6
Great Oxygenation Event, 19–20
Guantanamo Bay, 286, 304
Guillain–Barré syndrome, 129
Guy's Hospital, Sleep Disorders Centre, 14, 15–17, 20, 200

H1N1, 127–9
hallucination, 1, 11, 68, 114, 120, 171, 188
  auditory, 177
  Evelyn's, *see* Evelyn
  of homunculus, 192
  hypnagogic, 120, 121, 187, 189, 190, 261, 280
  and poor vision, 188–9
  sexual, 193
  of smell, 176
Hobson, Allan, 267–8, 270–1, 278–9, 280
human lymphocyte antigen (HLA), 126–7, 129
hypersexuality, 153, 227, 238, 239–40, 251
hypnagogic hallucination, 120, 121, 187, 189, 190, 261, 280
hypocretin, 117, 118, 120, 121, 122, 123, 124, 126, 127, 129, 140–1
hypothalamus, 21, 116, 123, 141, 251
  and H1N1, 128
  lateral, 117, 126, 262
  posterior, 116
hypoxia, 84–5
hysteria, 95

Inception, 284

incubi/succubi, 187, 192, 261–2
insomnia (*see also* sleep; sleep deprivation):
bedside worries concerning, 295
and bodily weight, 292, 293–4
Clair's problems with, *see* Claire
and clinical depression, 300
and cognitive behavioural therapy, 302–3, 304–5
and cognitive performance, 293
common condition, 290
David's problems with, *see* David
debilitating nature of, 12
deprivation conflated with, 292
and different sleep durations, 312–13
different types of, 290
genetic explanation for some of, 294–5
gradual worsening of, 296
and health risks, 294
and increased mortality, 292, 293
and medications, 301
and narcolepsy, 85, 140
numbers suffering, 3–4
obvious nature of, 30
paradoxical, 291
and schizophrenia, 300
and short sleep duration, 293
sleep deprivation to treat, 304
and suicidal thinking, 300
insular cortex, 177
Interpretation of Dreams (Freud), 308

Jackie (night activities), 36–9, 48–52, 55–6, 102, 200
and CPAP, 55
night driving by, 49–50, 51, 55
night motorcycling by, 38–9, 46, 48
and seeing/hearing activity, 48
Jalal, Baland, 192–3
Jamie (KLS), 232–41, 244–50, 256, 257
Janice (insular epilepsy), 165–8, 172–3, 176, 178–9, 180–1, 182–3
jet lag, 15
John (REM sleep behaviour disorder), 59–61, 65, 67, 69–71, 122, 188, 217
and day–night differences, 71, 72
and melatonin, 74–5
and physical strength, 72
and sleep apnoea, 73–4
Jok (Jamie's father), 233, 234, 235, 236, 237, 241, 244–5, 246, 247, 256, 257–8

Jouvet, Michel, 65, 67, 268

Katie (Alex's girlfriend), 45–6, 57
KGB, 287
Kim (wife of Phil), 131, 132–4, 135, 137–9
Kleine–Levin syndrome (KLS), 237–54, 256–7 (*see also* Jamie)
  age of onset of, 238
  charity that supports, 256–7
  conditions that mimic, 242
  many drugs tried for, 248–9
  and morbid hunger, 239
Kleine, Willi, 238, 239
Kleitman, Nathaniel, 18, 62, 63, 119
Klüver–Bucy syndrome, 242
Kryger, Meir, 267

lesions:
  correlation of, to symptoms or signs, 10
  and dreams, 11
  and Gage, 8–10
  as medical term for brain damage, 7–8
  sleep disorders as result of, 8
Levin, Max, 238, 239
levodopa, 148, 151, 152, 153, 156, 160
Lewy body disease, 68, 74
Linda (Robert's partner), 101–2, 103, 104–5, 107–9
lithium, 249
Liz (wife of John), 59–61, 65, 69–71, 72
  and melatonin, 74–5

McCartney, Paul, 270
mad-cow disease, 296
Magas of Cyrene, 82
Mammoth Cave, 18
The Man Who Mistook His Wife for a Hat (Sacks), 7
Maria (sleep apnoea), 77–80, 82, 83, 88–91, 93
Marie, Pierre, 96
Marks, Howard, 100
Mary Rose (RLS), 146–8, 149, 151–2, 153, 154, 156, 160–2
Megan (Jamie's girlfriend), 236, 246–7, 257
melatonin, 27–8, 30–1, 33, 56, 74
Mendeleev, Dmitri, 270
Mignot, Emmanuel, 254
migraine, 10–11, 85, 97, 154, 253, 254–6
  and electrical activity, 6
mind–body link, 96
Mitchison, Graeme, 278
'Mother's Little Helper', 301
Munchausen's, 95
  by proxy, 106–7

narcolepsy, 15, 67, 85, 98, 114–21 *passim*, 122, 123, 125–30 *passim*, 132, 135, 140–1, 190, 244, 251 (*see also* sleep)

and amygdala, 123–4
and battery-recharging, 283
cause of, 254
Christian's problems with, *see* Christian
in dogs, 116–17, 118
and H1N1, 127, 128
and hallucinations, 189
and insomnia, 85, 140
new treatments for, 140
and Pandemrix, 129
and sleep paralysis, 189
National Hospital for Neurology, 4
Nature, 62
Neurology and Neurosurgery Illustrated, 114, 140
New Yorker, 286
night terrors, 59–61, 63, 65 (*see also* dreams)
and non-REM parasomnias, 176
night-time activities, 36–54 (*see also* Alex; Jackie; sleepwalking)
and alcohol, 54
Alex's sleepwalking, 42–6
car driving, 49–50, 51, 55
motorbike ride, 38–9, 46
The Nightmare (Fuseli), 186–7
non-24-hour rhythm disorder, 25–6, 29
Nordlander, Nils, 155

Occam's razor, 10
olfactory bulb, 68
operculum, 178
Oriele (Jamie's mother), 233–4, 236, 244, 247, 257–8
orthosomnia, 311
Othello syndrome, 105–6

PANDAS, 235
Pandemrix, 128–9, 130
paradoxical kinesis, 73
parasomnia, 38, 42–3, 45, 46, 47, 53, 54, 56, 102–4, 176, 180, 200, 201, 202, 207, 209, 211–13, 222, 227, 301
and alcohol, 211, 213
and standard treatments, failure of, 229
parietal lobe, 171, 172, 177, 178
Parinaud, Henri, 96
Parkinson's disease, 68, 69, 71, 72–3, 74, 85, 116, 151, 153
Parks, Kenneth, 208–9
Pavlov, Ivan, 303
Penfield, Wilder, 172, 178
phantom limb, 192
Phil (cataplexy), 131–9, 261, 262, 283
The Pickwick Papers (Dickens), 81
pineal gland, 27, 28, 30–1, 32–3 (*see also* melatonin)
post-traumatic stress disorder (PTSD), 189, 214, 275, 276–7
and Asperger's, 213

prazosin, 277
primary motor cortex, 175
PTSD (post-traumatic stress disorder), 275

Ramachandran, V. S., 191, 192–3
rapid eye movement (REM) (*see also* sleep):
  in sleep and dreams, *see* dreams
  and sleep behaviour disorder, *see* sleep behaviour disorder
RBD, *see* REM sleep behaviour disorder
REM (rapid eye movement), *see* dreams
restless legs syndrome (RLS), 144–57, 229, 309 (*see also* Mary Rose)
  and drug dependency, 160, 162
  and heart problems and blood pressure, 160
  and iron, 156
  more common in women, 156–7
  numbers suffering, 3–4
  recognising, 162–3
  and sleep deprivation, 162
  and sleep-eating, 226–7 (*see also* sleep-eating)
retinohypothalamic tract, 22
Richards, Keith, 270
RLS (restless legs syndrome), 309
Robert, 99–105, 107–9
  and CPAP, 104–5
  and gaslighting, 107
Rolling Stones, 301
ropinirole, 151, 152, 153, 158, 160
Royal Observatory, Greenwich, 20–1

Sacks, Oliver, 7, 14, 116
Salk Institute, 2
Sanger Institute, 117
Sarah (partner of Tom), 198–201, 203, 205, 206–7, 212, 214–15, 216
'Satisfaction', 270
schizophrenia, 241, 273, 300
sensory cortex, 177
sexsomnia, 201, 202–3, 206–8, 211, 212, 214
Shelley, Mary, 269–70
sleep (*see also* circadian clock; insomnia; narcolepsy; night-time activities):
  and blinking and eyelid movements, 62–3
  and body's clock, 14, 17
  and brainwave fluctuations, 164
  in cats, 65
  as complex set of brain states, 2
  David's problems with, *see* David

delayed phase syndrome of, 17, 22–4 (*see also* Vincent (at Guys))
and delta waves, 40
deprivation of, 54
and dogs, 64
and dreams, *see* dreams
and earth's rotation, 17
epilepsy's links with, 173–4
extremes of, 39
and full bladder, 54
genes' influence on, 23
hallucination during, 184–5
and homeostatic mechanism, 13–14
and lesions, *see* lesions
and narcolepsy, 15, 67
non-REM, 13, 38, 40, 43, 46–7, 53, 54, 56, 103, 104, 120, 121, 174, 176, 189, 200–2 *passim*, 209, 222, 227, 269, 270, 301, 314 (*see also* sleep: REM)
and Occam's razor, 10
and out-of-body experiences, 191, 193, 261
paralysis during, 11, 65, 68, 85, 103–4, 114, 119, 120, 121, 122, 187–8, 189, 191–2, 193–4, 196–7
and phase syndromes, 22–4
preferences for timing of, 22–3
REM, 2, 13, 47, 62–6, 68, 73, 103–4, 119–22, 174, 187–8, 189, 196, 202, 262, 266–71, 272, 274–9 *passim*, 281, 282, 314 (*see also* sleep: non-REM)
REM sleep behaviour disorder (RBD), 65–9, 70–2, 73, 103, 202
sibling-shared, 142
and snoring, 11, 54, 55, 79, 80 (*see also* sleep apnoea)
stages of, 12–13
Tom's problems with, *see* Tom
tracking of, 311–12
uni-hemispheric, in animals, 39
sleep apnoea, 12, 73, 79–93 *passim*, 103–4, 143, 168, 189, 229
and Alzheimer's, 86
and calorie intake, 84–5
and dementia, 86
early references to, 82
and effects on brain, 88
and heart problems and blood pressure, 83
and neurology, 85
numbers suffering, 3–4, 81
treatments for, 89
and male–female differences, 71–2
and melatonin, 74
sleep deprivation, 12, 52, 54, 115, 121, 186, 190, 216, 286–8 (*see also* insomnia)
and appetite, 31

sleep deprivation – *continued*
and deeper sleep, 54
at Guantanamo, 286, 304
insomnia
conflated with, 292
and low oxygen, 85
and RLS, 162
and school
performance, 24, 30
to treat insomnia, 304
trigger of seizures, 86, 173
sleep-eating (*see also* Don):
and non-REM
sleep, 222, 227
only recently described, 220
and RLS, 226–7
and types of food
consumed, 221–2
sleep retraining, 304
sleep-talking:
and non-REM
parasomnias, 176
by Robert, 101–5
sleepwalking (*see also*
night-time activities; sleep):
by Alex, 42–7, 56–7
and brainwave activity, 40
and breathing concerns,
36
in childhood, 38, 54
and criminal acts,
208, 210–11
darker side of, 45
and environmental or
lifestyle factors, 53
and epilepsy, 41–2
and external factors, 211
and
'fright-fight-flight', 45, 47
and implanted
electrodes, 41
medicolegal history of, 208
and non-REM
parasomnias, 176
and open eyes, 66
questions concerning, 52–3
'quite extreme', 37
self-harm during, 37
sleep-talking a variation of,
see sleep-talking
and sound sleep, 54
and SPECT, 40–1, 42
and surgery, 41–2
treatments for, drug and
non-drug, 56–7
triggered by sounds, 53
and twins, 53
Socrates, 216
sodium oxybate, 136–7,
283
Sophie (Adrian's
daughter), 113–14
suprachiasmatic nucleus, 21–2,
27, 28, 31, 33
swine flu, 127

Tamiflu, 127
TATT (tired all the time),
76–7 (*see also* Maria)
temporal lobe, 123, 170,
171, 172, 177
thalamus, 244, 251

tired all the time
(TATT), 76–7
Tom (sleep problems),
198–201, 203–7, 212–
15, 216, 217
rape conviction of, 203–7,
212–13, 215
topiramate, 229
Tourette, Georges
Gilles de la, 96
twins, and sleepwalking, 53

untreatable sleepiness, 98–9

Vincent (at Guy's), 15–17, 22,
24–7, 28, 29, 32, 33–4, 35
and excellent exam
grades, 32
visual cortex, 6, 171, 172
von Economo,
Constantin, 116, 117

Walker, Matthew, 269, 275–6
Willis, Thomas, 144
World Health Organization,
30

'Yesterday', 270

Zeitgebers, 19, 22, 27, 29

DR. GUY LESCHZINER is a consultant neurologist and sleep physician, and is the clinical lead for one of the largest sleep services in Europe, based at Guy's Hospital in Central London. He sees patients with a range of sleep disorders, including narcolepsy, restless legs syndrome, sleep apnea, and nocturnal epilepsy, and is actively involved in research and teaching. He is also the presenter of the BBC radio series *Mysteries of Sleep*.